AN ESSENCE OF BIOTECHNOLOGY

AN ESSENCE OF BIOTECHNOLOGY

Editors

Dr. Sumedha Mehta Mohan • Chintanu Kumar Sarmah

Publications

KALPAZ PUBLICATIONS

DELHI-110052

An Essence of Biotechnology

Rs. 690

ISBN: 81-7835-469-1

Published in 2005 in India by
Kalpaz Publications,
C-30, Satyawati Nagar, Delhi-110052
E-mail : Kalpaz@hotmail.com

Laser Typesetting at : Master Computers, Delhi
Printed at : Salasar Imaging System, Delhi

Contents

Preface

Biotechnology is a general term describing the directed modification of biological processes. This may be accomplished by introducing new genes into organisms, breeding organisms to form new variants, or treating organisms with specific compounds. Biotechnology-derived products have the potential to positively change our society, from improving health care and increasing agricultural products to producing a cleaner environment. The public markets recognise this potential, and a lot of money has flowed into biotechnology companies. Developments such as the sequencing of the human genome provide a foundation that will support decades of research and discovery.

Biotechnology covers a multitude of different applications : traditional biotechnology which refers to the conventional techniques that have been used for many centuries to produce beer, wine, cheese and many other foods, while modern biotechnology embraces all methods of genetic modification by recombinant DNA and cell fusion techniques to yield new drugs or to introduce new trades into commercial crops and animals as examples.

The present work facilitates an insight into a number of core and important areas of biotechnology. All facts and figures have been carefully and authoritatively collected and nothing has been left to chance or conjecture. We have consulted a large number of published works, drawn information from them and thus owe a depth of gratitude to all such publishers, authors, editors and compilers. The credit for producing and furnishing us authentic data actually goes to them. We have only weaved and consolidated the details collected from them.

Traditional biotechnology has been in practice since the beginning of history records. But recent developments in molecular biology have given biotechnology new meaning, new prominence and new potential. Modern biotechnology has a dramatic effect on the world economy and society. Genetic engineering, recombinant DNA technology and bioinformatics are a few examples of modern day biotechnology. Through genetic engineering, genetically modified organisms are formed. Genes introduced may make the organism more resistant to disease, may influence the rate of fruit ripening or may increase productivity.

Recombinant DNA technology is the laboratory manipulation of DNA in which DNA or fragments of DNA from different sources are cut and recombined using enzymes. This rDNA is then inserted into a living organism. This technology allows researchers to move genetic information between unrelated organisms to produce desired products or characteristics or to eliminate undesirable characteristics.

Modern biotechnology offers effective techniques to improve product quality, nutritional content and economic benefits. The genetic makeup of plants and animals can be modified by either insertion of new useful genes or removal of unwanted ones. Biotechnology is changing the way plants and animals are used boosting their value to growers, processors and consumers.

Editors

1

A Gentle Introduction to a Big Idea

The discovery of the structure of DNA in 1953 and the identification of DNA as the genetic material in all life allowed for great leaps in our understanding of life-forms from bacteria to plants to humans. Armed with a better understanding of how organisms 'work', we are now able to engineer life to suit our needs.

Biotechnology is a general term describing the directed modification of biological processes. This may be accomplished by introducing new genes into organisms, breeding organisms to form new variants, or treating organisms with specific compounds. Biomedicine is medicine based upon the principles of biology and biochemistry-the application of biotechnology to medicine.

Biotechnology-derived products have the potential to positively change our society, from improving health care and increasing agricultural products to producing a cleaner environment. The public markets recognize this potential, and a lot of money has flowed into biotechnology companies. Sales, revenues and market capitalization all more than doubled during the period from 1993 to 1999. The year 2000 saw more money invested in biotechnology firms than the previous five years combined. The business models have matured and the tools of biotechnology research are refined to the extent that the industry is ready for unprecedented dramatic growth.

Developments such as the sequencing of the human genome provide a foundation that will support decades of research and discovery. An important issue in the biotechnology industry is intellectual property-some argue that the scope of patent protection actually defines the industry. Because many biotechnology companies must partner with larger firms to complete product development, merger and acquisition activity is monitored to gauge the maturity of the industry. Amgen and Genentech two of the biggest biotech companies, but scores of smaller companies such as Biogen, Human Genome Sciences, Incyte, and Millenium Pharmaceuticals have aggressive research programs that could potentially produce products to bring them to the top.

Development of Biotechnology

Long before the term "biotechnology" was coined for the process of using living organisms to produce improved commodities, people were utilizing living micro-organisms to produce valuable products. A list of early biotechnology applications is give in table 1.

Table 1. Early biotechnology applications

Proving bread with leaven	prehistoric period
Fermentation of juices to alcoholic beverages	prehistoric period
Knowledge of vinegar formation from fermented juices	prehistoric period
Cultivation of vine	before 2000 BC
Manufacture of beer in Babylonia and Egypt	3rd century BC
Wine growing promoted by Roman Emperor Marcus Aurelius Probus	3rd century AD

Production of spirits of wine (ethanol)	1150
Vinegar manufacturing industry	14th century
Discovery of the fermentation properties of yeast by Erxleben	1818
Description of lactic acid fermentation by Pasteur	1857
Detection of fermentation enzymes in yeast by Buchner	1897
Discovery of penicillin by Fleming	1928/29
Discovery of many other antibiotics	from about 1945

Since then "biotechnology" has rapidly progressed and expanded. In the mid-forties, scale-up and commercial production of antibiotics such as penicillin occurred. The techniques used were:

(a) isolation of an organism producing the chemical of interest using screening/selection procedures, and

(b) improvement of production yields via mutagenesis of the organism or optimization of media and fermentation conditions.

This type of "antique" biotechnology is limited to chemicals produced in nature. It is also limited by its trial-and-error approach, and requires a lengthy timeframe for yield improvement.

About two decades ago, biotechnology became much more of a science rather than an art. Regions of DNA were found to contain information that would lead to synthesis of specific proteins. Each of these proteins have their own identity and function; many catalyze chemical reactions, and others are structural components of entities in cells.

If one now is able to express a natural gene in simple bacteria such as *Escherichia coli* (*E. coli*), a bacterium living in intestines that has become the model organism for much of biotechnology, one can have this bacterium make a lot of the protein coded for by the gene, regardless its source. The techniques used for this development include

(a) isolation of the gene coding for a protein of interest,

(b) cloning of this gene into an appropriate production host, and

(c) improving expression by using better promoters, tighter regulation, etc.; together these techniques are known as recombinant DNA techniques.

The commercial implications are that a large number of proteins, existing only in tiny quantities in nature, can now be mass-produced if needed. Also, the yields of (bio)chemicals to be produced can be increased much faster than was possible with classical fermentation.

These modern biotechnology techniques started with the expression of human genes such as that coding for insulin, but have since been extended to mammalian, microbial, and plant genes. Also, the spectrum of "bioreactors" (organisms used for production) recently has been broadened to include a variety of animals and plants. Perceived needs and marketability, the researchers' imagination, ethics, and governmental regulations essentially are the major factors in setting the stage and boundaries for developments in biotechnology.

About a decade ago, "protein engineering" became possible as an offshoot of the recombinant DNA technology. Protein engineering differs from "classical" biotechnology in that it is concerned with producing new (man-made) proteins which have been modified or improved in some way. The techniques involved in protein engineering are more complicated than before, and involve:

(a) various types of mutagenesis (to cause changes in specific locations or regions of a gene to produce a new gene product),

(b) expression of the new gene to form a stable protein,

(c) characterization of the structure and function of the protein produced, and

(d) selection of new locations or regions to modify as a result of this characterization.

In the mid-eighties and early-nineties, it has become possible to transform (genetically modify) plants and animals that are important for food production. "Transgenic" animals and plants, including cows, sheep, tomatoes, tobacco, potato, and cotton have now been obtained. Genes introduced may make the organism more resistant to disease, may influence the rate of fruit ripening, or may increase productivity. As this approach leads to release of genetically altered organisms into the environment, this part of biotechnology is quite strictly regulated at government levels.

An overview of recombinant DNA based biotechnology is given in table 2.

Table 2. Recombinant DNA based biotechnology

1953	Double helix structure of DNA is first described by Watson and Crick.
1973	Cohen and Boyer develop genetic engineering techniques to "cut and paste" DNA and to amplify the new DNA in bacteria.
1977	The first human protein (somatostatin) is produced in a bacterium (E. coli).
1982	The first recombinant protein (human insulin) appears on the market.
1983	Polymerase chain reaction (PCR) technique conceived.
1990	Launch of the Human Genome Project (HGP), an international effort to sequence the human genome.
1995	The first genome sequence of an organism (Haemophilus influenzae) is determined.
2000	A first draft of the human genome sequence is completed.

Sub-Fields of Biotechnology

There are a number of jargon terms for sub-fields of biotech.

— ***Red biotechnology*** is biotechnology applied to medical processes. Some examples are the designing of organisms to produce antibiotics, and the engineering of genetic cures to diseases through genomic manipulation.

— ***White biotechnology***, also known as grey biotechnology, is biotechnology applied to industrial processes. An example is the designing of an organism to produce a useful chemical. White biotechnology tends to consume less in resources that traditional processes when used to produce industrial goods.

— ***Green biotechnology*** is biotechnology applied to agricultural processes. An example is the designing of an organism to grow under specific environmental conditions or in the presence or absence of certain agricultural chemicals. Green biotechnology tends to produce environmentally more friendly solutions than traditional industrial agriculture. An example of this is the engineering of a plant to express a pesticide, thereby eliminating the need for external application of pesticides.

— ***Bioinformatics*** is a an interdisciplinary field which addresses biological problems using computational techniques. The field is also often referred to as computational biology. It plays a key role in various areas like functional genomics, structural genomics, and proteomics amongst others, and forms a key component in biotechnology and pharmaceutical sector.

— The term ***blue biotechnology*** has also been used to describe the marine and aquatic applications of biotechnology, but its use is relatively rare.

Modern Biotechnology

Traditional biotechnology has been practiced since the beginning of history. It has been used to: bake bread, brew alcoholic beverages, and breed food crops or domestic animals. But recent developments in molecular biology have given biotechnology new meaning, new prominence, and new potential. It is modern biotechnology that has captured the attention of the public. Modern biotechnology can have a dramatic effect on the world economy and society.

One example of modern biotechnology is genetic engineering. Genetic engineering is the process of transferring individual genes between organisms or modifying the genes in an organism to remove or add a desired trait or characteristic. Examples of genetic engineering are described later in this document. Through genetic engineering, genetically modified crops or organisms are formed. These GM crops or GMOs are used to produce biotech-derived foods. It is this specific type of modern biotechnology, genetic engineering, that seems to generate the most attention and concern by consumers and consumer groups. What is interesting is that modern biotechnology is far more precise than traditional forms of biotechnology and so is viewed by some as being far safer.

Principles of Modern Biotechnology

All organisms are made up of cells that are programmed by the same basic genetic material, called DNA (deoxyribonucleic acid). Each unit of DNA is made up of a combination of the following nucleotides—adenine (A), guanine (G), thymine (T), and cytosine (C)—as well as a sugar and a phosphate. These nucleotides pair up into strands that twist together into a spiral structure call a "double helix." This double helix is DNA. Segments of the DNA tell individual cells how to produce specific proteins. These segments are genes. It is the presence or absence of the specific protein that gives an organism a trait or characteristic.

More than 10,000 different genes are found in most plant and animal species. This total set of genes for an organism is organized into chromosomes within the cell nucleus. The process by which a multicellular organism develops from a single cell through an embryo stage into an adult is ultimately controlled by the genetic information of the cell, as well as interaction of genes and gene products with environmental factors.

When cells reproduce, the DNA strands of the double helix separate. Because nucleotide A always pairs with T and G always pairs with C, each DNA strand serves as a precise blueprint for a specific protein. Except for mutations or mistakes in the replication process, a single cell is equipped with the information to replicate into millions of identical cells. Because all organisms are made up of the same type of genetic material (nucleotides A, T, G, and C), biotechnologists use enzymes to cut and remove DNA segments from one organism and recombine it with DNA in another organism. This is called recombinant DNA (rDNA) technology, and it is one of the basic tools of modern biotechnology.

rDNA technology is the laboratory manipulation of DNA in which DNA, or fragments of DNA from different sources, are cut and recombined using enzymes. This recombinant DNA is then inserted into a living organism. rDNA technology is usually used synonymously with genetic engineering. rDNA technology allows researchers to move genetic information between unrelated organisms to produce desired products or characteristics or to eliminate undesirable characteristics.

Genetic engineering is the technique of removing, modifying or adding genes to a DNA molecule in order to change the information it contains. By changing this information, genetic engineering changes the type or amount of proteins an organism is capable of producing.

Genetic engineering is used in the production of drugs, human gene therapy, and the development of improved plants. For example, an "insect protection" gene *Bacillus thuringiensis* (*Bt*) has been inserted into several crops—corn, cotton, and potatoes—to give farmers new tools for integrated pest management. *Bt* corn is resistant to European corn borer. This inherent resistance thus reduces a farmer's pesticide use for controlling European corn borer as well as enhances the crop yield.

Although major genetic improvements have been made in crops, progress in conventional breeding programs has been slow. In fact, most crops produce less than their full genetic potential. These shortfalls in yield are due to the inability of crops to tolerate or adapt to environmental stress, pests, and diseases.

For example, some of the world's highest yields of potatoes are in Idaho under irrigation, but in 1993 both quality and yield were severely reduced because of cold, wet weather and widespread frost damage during June. Some of the world's best bread wheats and malting barleys are produced in the north-central states, but in 1993 the disease Fusarium caused an estimated $1 billion in damage.

Scientists have the ability to insert genes that give biological defense against diseases and insects, thus reducing the need for chemical pesticides, and they will soon be able to convey genetic traits that enable crops to better withstand harsh conditions, such as drought. The International Laboratory for Tropical Agricultural Biotechnology (ILTAB) is developing transformation techniques and applications for control of diseases caused by plant viruses in tropical plants such as rice, cassava and tomato.

In 1995, ILTAB reported the first transfer through biotechnology of a resistance gene from a wild species of rice to a susceptible cultivated rice variety. The transferred gene expressed resistance to *Xanthomonas oryzae*, a bacterium which can destroy the crop through disease. The resistant gene was transferred into susceptible rice varieties that are cultivated on more than 24 million hectares around the world.

Benefits can also be seen in the environment, where insect-protected biotech crops reduce the need for chemical pesticide use. Insect-protected crops allow for less potential exposure of farms and groundwater to chemical residues, while providing farmers with season-long control. Also by reducing the need for pest control, impacts and resources spent on the land are less, thereby preserving the topsoil.

Major advances also have been made through conventional breeding and selection of livestock, but significant gains can still be made by using biotechnology. Currently, farmers in the U.S. spend $17 billion on animal health. Diseases such as hog cholera and pests such as screwworm have been eradicated. Uses of biotechnology in animal production include development of vaccines to protect animals from disease, production of several calves from one embryo (cloning), increase of animal growth rate, and rapid disease detection.

Applications of Modern Biotechnology

Modern biotechnology has offered opportunities to produce more nutritious and better tasting foods, higher crop yields and plants that are naturally protected from disease and insects. Modern biotechnology allows for the transfer of only one or a few desirable genes, thereby permitting scientists to develop crops with specific beneficial traits and reduce undesirable traits. Traditional biotechnology such as cross-pollination in corn produces numerous, non-selective changes.

Genetic modifications have produced fruits that can ripen on the vine for better taste, yet have longer shelf lives through delayed pectin degradation. Tomatoes and other produce containing increased levels of certain nutrients, such as vitamin C, vitamin E, and or beta carotene, help protect against the risk of chronic diseases, such as some cancers and heart disease. Similarly introducing genes that increase available iron levels in rice three-fold is a potential remedy for iron deficiency, a condition that effects more than two billion people and causes anemia in about half that number. Most of the today's hard cheese products are made with a biotech enzyme called chymosin. This is produced by genetically engineered bacteria which is considered more purer and plentiful than it's naturally occurring counterpart, rennet, which is derived from calf stomach tissue.

In 1992, Monsanto Company successfully inserted a gene from a bacterium into the Russet Burbank potato. This gene increases the starch content of the potato. Higher starch content reduces oil absorption during frying, thereby lowering the cost of processing french fries and chips and reducing the fat content in the finished product. This product is still awaiting final development and approval.

Modern biotechnology offers effective techniques to address food safety concerns. Biotechnical methods may be used to decrease the time necessary to detect foodborne pathogens, toxins, and chemical contaminants, as well as to increase detection sensitivity. Enzymes, antibodies, and microorganisms produced using rDNA techniques are being used to monitor food production and processing systems for quality control.

Biotechnology can compress the time frame required to translate fundamental discoveries into applications. This is done by controlling which genes are altered in an organized fashion. For example, a known gene sequence from a corn plant can be altered to improve yield, increase drought tolerance, and produce insect resistance (*Bt*) in one generation. Conventional breeding techniques would take several years and would require that a field of corn is grown and each trait is selected from individual stalks of corn. The ears of corn from selected stalks with each desired trait (e.g, drought tolerance and yield performance) would then be grown and combined (cross-pollinated). Their offspring (hybrid) would be further selected for the desired result (a high performing corn with drought tolerance). With improved technology and knowledge about agricultural organisms, processes, and ecosystems, opportunities will emerge to produce new and improved agricultural products in an environmentally sound manner.

Modern biotechnology offers opportunities to improve product quality, nutritional content, and economic benefits. The genetic makeup of plants and animals can be modified by either insertion of new useful genes or removal of unwanted ones. Biotechnology is changing the way plants and animals are grown, boosting their value to growers, processors, and consumers.

Human applications

Biotechnical methods are now used to produce many proteins for pharmaceutical and other specialized purposes. A harmless strain of *Escherichia coli* bacteria, given a copy of the gene for human insulin, can make insulin. As these genetically modified (GM) bacterial cells age, they produce human insulin, which can be purified and used to treat diabetes in humans. Microorganisms can also be modified to produce digestive enzymes.

In the future, these microorganisms could be colonized in the intestinal tract of persons with digestive enzyme insufficiencies. Products of modern biotechnology include artificial blood vessels from collagen tubes coated with a layer of the anticoagulant heparin. Gene therapy–altering DNA within cells in an organism to treat or cure a disease–is one of the most promising areas of biotechnology research. New genetic therapies are being developed to treat diseases such as cystic fibrosis, AIDS and cancer.

DNA fingerprinting is the process of cross matching two strands of DNA. In criminal investigations, DNA from samples of hair, bodily fluids or skin at a crime scene are compared with those obtained from the suspects. In practice, it has become one of the most powerful and widely known applications of biotechnology today. Another process, polymerase chain reaction (PCR), is also being used to more quickly and accurately identify the presence of infections such as AIDS, Lyme disease and Chlamydia.

Paternity determination is possible because a child's DNA pattern is inherited, half from the mother and half from the father. To establish paternity, DNA fingerprints of the mother, child and the alleged father are compared. The matching sequences of the mother and the child are eliminated from the child's DNA fingerprint; what remains comes from the biological father. These segments are then compared for a match with the DNA fingerprint of the alleged father.

DNA testing is also used on human fossils to determine how closely related fossil samples are from different geographic locations and geologic areas. The results shed light on the history of human evolution and the manner in which human ancestors settled different parts of the world.

Industrial applications

Industrial biotechnology applies the techniques of modern molecular biology to improve the efficiency and reduce the environmental impacts of industrial processes like textile, paper and pulp, and chemical manufacturing. For example, industrial biotechnology companies develop biocatalysts, such as enzymes, to synthesize chemicals. Enzymes are proteins produced by all organisms. Using biotechnology, the desired enzyme can be manufactured in commercial quantities.

Commodity chemicals (e.g., polymer-grade acrylamide) and speciality chemicals can be produced using biotech applications. Traditional chemical synthesis involves large amounts of energy and often-undesirable products, such as HCl. Using biocatalysts, the same chemicals can be produced more economically and more environmentally friendly. An example would be the substitution of protease in detergents for other cleaning compounds.

Detergent proteases, which remove protein impurities, are essential components of modern detergents. They are used to break down protein, starch, and fatty acids present on items being washed. Protease production results in a biomass that in turn yields a useful byproduct-an organic fertilizer. Biotechnology is also used in the textile industry for the finishing of fabrics and garments. Biotechnology also produces biotech-derived cotton that is warmer, stronger, has improved dye uptake and retention, enhanced absorbency, and wrinkle-and shrink-resistance.

Some agricultural crops, such as corn, can be used in place of petroleum to produce chemicals. The crop's sugar can be fermented to acid, which can be then used as an intermediate to produce other chemical feedstocks for various products. It has been projected that 30% of the world's chemical and fuel needs could be supplied by such renewable resources in the first half of the next century. It has been demonstrated, at test scale, that biopulping reduces the electrical energy required for wood pulping process by 30%.

Environmental applications

Environmental biotechnology is used in waste treatment and pollution prevention. Environmental biotechnology can more efficiently clean up many wastes than conventional methods and greatly reduce our dependence on methods for land-based disposal.

Every organism ingests nutrients to live and produces by-products as a result. Different organisms need different types of nutrients. Some bacteria thrive on the chemical components of waste products. Environmental engineers use bioremediation, the broadest application of environmental biotechnology, in two basic ways. They introduce nutrients to stimulate the activity of bacteria already present in the soil at a waste site, or add new bacteria to the soil. The bacteria digest the waste at the site and turn it into harmless byproducts. After the bacteria consume the waste materials, they die off or return to their normal population levels in the environment.

Bioremediation, is an area of increasing interest. Through application of biotechnical methods, enzyme bioreactors are being developed that will pretreat some industrial waste and food waste components and allow their removal through the sewage system rather than through solid waste disposal mechanisms. Waste can also be converted to biofuel to run generators. Microbes can be induced to produce enzymes needed to convert plant and vegetable materials into building blocks for biodegradable plastics.

In some cases, the byproducts of the pollution-fighting microorganisms are themselves useful. For example, methane can be derived from a form of bacteria that degrades sulfur liquor, a waste product of paper manufacturing. This methane can then be used as a fuel or in other industrial processes.

Applications to food and beverage industry

Survey research over the past decade shows that biotechnology is not likely to become an important issue for most food consumers. Consumers find biotechnology acceptable when they believe it offers benefits and it is safe. Surveys have consistently found that a majority of consumers are willing to buy insect-protected food crops developed through biotechnology that use fewer chemical pesticides, as well as more nutritious foods. American consumers also appreciate the role that biotechnology can play in feeding the world. Research shows that consumers are much less supportive of all biotechnology applications.

Surveys conducted since 1992 in US show that relatively few consumers have heard or read much about biotechnology. News about the cloned sheep pushed awareness to 50 percent in March 1997. Surveys in the first three months of 2000 show that awareness has fallen back to just over one-third in the United States. Such trends reflect the fact that most people get their information about biotechnology from the media. Unfortunately, many consumers also do not understand some fundamental principles of biology. European consumer awareness is somewhat higher, but knowledge is still low.

Consumers look to health professionals and scientific experts for credible information, but place relatively little trust in the activists who oppose biotechnology. Research shows that acceptance increases significantly when American consumers learn that organizations such as the National Academy of Sciences and the U.S. Food and Drug Administration have determined that biotech-derived foods are safe.

In contrast, European consumers express the most trust in those groups that oppose biotechnology. They have much less confidence in government, industry, or even scientists. American culture is more supportive and rewarding of new technology. Europeans tend to view food differently from U.S. consumers. In fact, some Europeans reject all American food products. Europeans also want to protect their small farms to maintain open space and rural employment. Such forces underlie much of the European anxiety about agricultural biotechnology.

Most of the industry leaders interviewed are quite enthusiastic about the benefits of biotechnology—

especially in terms of increased food availability, enhanced nutrition, and environmental protection. Most feel that biotechnology has already provided benefits to consumers. Almost all recognize that foods developed through biotechnology have already been part of consumers' everyday diet. They clearly do not agree with most of the opponents' claims and tend to have almost no trust in such groups.

Their main concerns involve lack of consumer acceptance—not the safety of the foods. They express high levels of confidence in the science and the regulatory process. In fact, almost none feel that biotechnology should not be used because of uncertain, potential risks. Most food industry leaders do not feel it is necessary to have special labels on biotech-derived foods. They express concerns that such labels would be perceived as a warning by consumers. They also worried that the need to segregate commodities would pose financial and logistical burdens on everyone in the system—including consumers. Food industry leaders recognize a major need to educate the public about biotechnology. They look to third parties, such as university and government scientists to provide such leadership.

Research shows that consumers will accept biotech foods if they see a benefit to themselves or society and if the price is right. Their responses to foods developed through biotechnology are basically the same as for any other food—taste, nutrition, price, safety and convenience are the major factors that influence our decisions about which foods to eat. How seeds and food ingredients are developed will only be relevant for a relatively small group of concerned, consumers.

The potential for public concerns has led several food companies to change their products to avoid biotech-derived ingredients. For example, Gerber Foods received threats from Greenpeace because they had determined the company was using biotech-derived food ingredients.

The company firmly believes that the biotech foods are safe to consume. Gerber agreed to drop some of its existing corn and soybean suppliers in favor of ones that can produce crops that are not genetically altered. It became an issue that is suddenly confronting other food companies. A private manufacturer in California, called Healthy Times Natural Food has switched from Canola oil to safflower oil after facing questions from Greenpeace. The controversy is due in part to the fact that the organic industry is using public concern as a tool for marketing their products as free of biotech ingredients.

Biotechnology and the Developing World

Experts anticipate the world's population in 2050 to be approximately 8.7 billion persons. The world's population is growing, but its surface area is not. Compounding the effects of population growth is the fact that most of the earth's ideal farming land is already being utilized. To avoid damaging environmentally sensitive areas, such as rain forests, we need to increase crop yields for land currently in use. By increasing crop yields, through the use of biotechnology the constant need to clear more land for growing food is reduced.

Countries in Asia, Africa, and elsewhere are grappling with how to continue feeding a growing population. They are also trying to benefit more from their existing resources. Biotechnology holds the key to increasing the yield of staple crops by allowing farmers to reap bigger harvests from currently cultivated land, while preserving the land's ability to support continued farming.

Malnutrition in underdeveloped countries is also being combated with biotechnology. Vitamin A deficiency is a common problem in poor countries. Rice breeders are using biotechnology to build Vitamin A into the rice. The Rockefeller Foundation is sponsoring research on "golden rice", a crop designed to improve nutrition in the developing world. A second phase of the project will increase the iron content in rice to combat anaemia, which is widespread problem among women and children in underdeveloped countries. Golden rice, expected to be for sale in Asia in less than five years, will offer dramatic improvements in nutrition and health for millions of people, with little additional costs to consumers.

Similar initiatives using genetic manipulation are aimed at making crops more productive by reducing their dependence on pesticides, fertilizers and irrigation, or by increasing their resistance to plant diseases. Increased crop yield, greater flexibility in growing environments, less use of chemical pesticides and improved nutritional content make agricultural biotechnology, quite literally, the future of the world's food supply.

As biotechnology has become widely used, questions and concerns have also been raised. The most vocal opposition has come from European countries. One of the main areas of concern is the safety of genetically engineered food In assessing the benefits and risks involved in the use of modern biotechnology, there are a series of issues to be addressed so that informed decisions can be made.

In making value judgments about risks and benefits in the use of biotechnology, it is important to distinguish between technology-inherent risks and technology-transcending risks. The former includes assessing any risks associated with food safety and the behavior of a biotechnology-based product in the environment. The latter involve the political and social context in which the technology is used, including how these uses may benefit or harm the interests of different groups in society.

The health effects of foods grown from genetically engineered crop depend on the composition of the food itself. Any new product may have either beneficial or occasional harmful effects on human health. For example, a biotech-derived food with a higher content of digestible iron is likely to have a positive effect if consumed by iron-deficient individuals. Alternatively, the transfer of genes from one species to another may also transfer the risk for exposure to allergens. These risks are systematically evaluated by food administration agencies and identified prior to commercialization. Individuals allergic to certain nuts, for example, need to know if genes conveying this trait are transferred to other foods such as soybeans. Labeling would be required if such crops were available to consumers.

Among the potential ecological risks identified are increased weediness, due to cross-pollination from genetically modified crops spreads to other plants in nearby fields. This may allow the spread of traits such as herbicide-resistance to non-target plants that could potentially develop into weeds. This ecological risk is assessed when deciding if a plant with a given trait should be released into a particular environment, and if so, under what conditions.

Other potential ecological risks stem from the use of genetically modified corn and cotton with insecticidal genes from *Bacillus thuringiensis* (*Bt*) genes. This may lead to the development of resistance to *Bt* in insect populations exposed to the biotech-derived crop. There also may be risks to non-target species, such as birds and butterflies, from the plants with *Bt* genes. The monitoring of these effects of new crops in the environment and implementation of effective risk management approaches is an essential component of further research. It is also important to keep all risks in perspective by comparing the products of biotechnology and conventional agriculture.

The reduction of biodiversity would represent a technology-transcending risk. Reduced biological diversity due to destruction of tropical forests, conversion of land to agriculture, overfishing, and the other practices to feed a growing world population is a significant loss far more than any potential loss of biodiversity due to biotech-derived crop varieties. Improved governance and international support are necessary to limit loss of biodiversity.

National governments and international policy making bodies rely on food scientists and others to develop innovations that will create marketable food products and increase food supplies. Governments also rely on scientific research because they are responsible for setting health and safety standards regarding new developments. International organizations can suggest policy approaches and help develop international treaties that are ratified by national governments.

Economic success in the competitive international market demands that food production become more efficient and profitable. National governments and international organizations support food biotechnology as a means to avoid global food shortages. Many policy making bodies are also trying to balance support of the food biotechnology industry with public calls for their regulation. Such regulations are necessary to protect public health and safety, to promote international trade, conserve natural resources, and account for ethical issues.

The majority of processed foods on the market contain soy or corn ingredients that come from GM plants. To date none have posed a food safety risk. The chief safety concerns are the potential to alter nutrient content or introduce allergens. Federal agencies involved in biotechnology regulation include the U.S. Department of Agriculture (USDA) which evaluates agricultural production processes for all foods; the Food and Drug Administration (FDA), which evaluates whole non-animal foods (seafood), food ingredients, and food additives; and the Environmental Protection Agency (EPA), which evaluates plants with insecticidal properties.

Developers of GM plants and biotech-derived foods are required to consult with FDA prior to the commercialization of the product. This consultation procedure entails a science-based safety assessment of the product that focuses on the protection of the consumer, developer, and the environment. Thus developers, have a strong incentive to cooperate fully with FDA and the other agencies prior to marketing their products.

Labeling food derived from GM plants and animals is an important, but complex issue. Some consumers and consumer groups believe they have a right to know whether biotechnology was used to produce food. Others believe labeling is not necessary if foods are essentially equivalent in composition. Regulatory agencies are concerned with ensuring that food labels are both true and not misleading. In accordance with their statutory mandate, the FDA has determined that a food product should be labeled as a product of biotechnology only if it has been changed in some significant way.

Based on the input from three public hearings, the FDA has recognized that the most effective way to allow for informed choice will be a system of voluntary labeling for foods not produced through biotechnology. If the demand for "GM-free" food is real, a market will become viable. In this case, meaningful choice can be provided to concerned consumers without imposing costs on or denying benefits to the majority of consumers, who support biotechnology. This may be the only viable and cost-effective option for the food industry.

National surveys conducted with American consumers between 1997 and 2000 found about three-quarters of consumers supported this FDA labeling policy. There is evidence from focus groups that US consumers are already overwhelmed by the level of detail on food labels and mainly look for relevant information about nutrition. Recent focus groups have demonstrated that the wording on labels has a significant effect on consumer understanding and acceptance of biotechnology. Any label messages need to be simple and clear. However, labeling will be counterproductive without education.

Requirements for all food labels mandate proper identification of products and notice of health or safety concerns. Labeling would be required in the following instances—but not simply because the products were made using biotechnology.

— Potential food allergy is an example of a health or safety risk that would mandate a product label (e.g., if a gene has been transferred from a known allergen, such as a peanut).

— When substantial changes have been made to the expected nutritional composition of a food (e.g., vitamins have been added to a particular type of produce).

— If the common identity of a food product has been changed (e.g., if the composition of a fruit or vegetable becomes significantly different it would have to have a different name).

Labeling of biotech-derived foods will be required in some countries for cultural and religious reasons. These countries also believe labeling should be required simply because the consumers want to know more about the content of the food or how it was produced. Some have suggested that such labeling initiatives have been driven by economic and political interests rather than by scientific need. Consumers in these countries will ultimately end up paying even higher prices for their food as a result of such strict labeling policies.

The applications of biotechnology are so broad, and the advantages so compelling, that virtually every industry is using this technology. Developments are underway in areas as diverse as pharmaceuticals, diagnostics, textiles, aquaculture, forestry, chemicals, household products, environmental cleanup, food processing and forensics to name a few. Biotechnology is enabling these industries to make new or better products, often with greater speed, efficiency and flexibility.

Biotechnology holds significant promise to the future but certain amount of risk is associated with any area. Biotechnology must continue to be carefully regulated so that the maximum benefits are received with the least risk.

Biotechnology is at a crossroads in terms of public acceptance. Many countires have not yet formed a solid opinion on this complex issue. International developments over the next few years will certainly have a major influence on the long-term viability of biotechnology. The future of the world food supply depends upon how well scientists, government, and the food industry are able to communicate with consumers about the benefits and safety of the technology.

References

1. Applications of Biotechnology to Crops: Benefits and Risks. *Council for Agricultural Science and Technology, Issue Paper, Number 12, Dec. 1999.*
2. *Bruhn CM (1992)* Consumer concerns and educational strategies: focus on biotechnology, *Food Technology, 46(3):80-102.*
3. Food—people who would have the most to lose. *New York Times Editorial, November 19, 1999.*
4. *Gates B (1996)* The Business of Biotechnology. *New York Times, June 18.*
5. Genetic Engineering and Society, Arizona State University, USA.
6. Genetically Engineered Foods, Fears and Facts: FDA. *Consumer 27(1). January/February 1993, pp11-14.*
7. *Hoban TJ, Kendall PA (1993)* Consumer Attitudes about Food Biotechnology. *Project Report 1993, Raleigh, North Carolina State University and Colorado State University; pp 1-36.*
8. *Keener KM, Hoban TJ and Balasubramanian RK (2000)* Biotechnology and Its Applications. *In Review. Distributed at Eloise S. Cofer Symposium on Biotechnology on October 27, 2000 in Research Triangle Park.*
9. Myths and Facts about Food Biotechnology. *Food Insight, September/October 1999.*

2

Redesigning our Inevitable Genetic Makeup

Genetic engineering is the reshuffling of genes usually from one species to another. It often involves the isolation, manipulation and reintroduction of DNA into model organisms, usually to express a protein. The aim is to introduce new characteristics to an organism in order to increase its usefulness, such as increasing the yield of a crop species, introducing a novel characteristic, or producing a new protein or enzyme. Examples are the production of human insulin through the use of modified bacteria and the production of new types of experimental mice such as the OncoMouse (cancer mouse) for research, through genetic redesign. To understand genetic engineering it is best to start with some basic cell biology.

Basics of Cell Biology

A cell is the smallest living unit, the basic structural and functional unit of all living matter, whether that is a plant, an animal or a fungus. Some organisms such as amoebae, bacteria, some algae and fungi are single-celled—the entire organism is contained in just one cell.

Humans are quite different and are made up of approximately 3 million cells. Cells can take many shapes depending on their function, but commonly they will look like a brick with rounded comers or an angular blob—a building block.Cells are stacked together to make up tissues, organs or structures (brain, liver, bones, skin, leaves, fruit etc.).

In an organism, cells depend on each other to perform various functions and tasks; some cells will produce enzymes, others will store sugars or fat; different cells again will build the skeleton or be in charge of communication like nerve cells; others are there for defence, such as white blood cells or stinging cells in jelly fish and plants. In order to be a fully functional part of the whole, most cells have got the same information and resources and the same basic equipment.

A cell belonging to higher organisms (e.g. plant or animal) is composed of:

— a cell membrane enclosing the whole cell. (Plant cells have an additional cell wall for structural reinforcement.)

— many organelles, which are functional components equivalent to the organs in the body of an animal e.g. for digestion, storage, excretion.

— a nucleus, the command centre of the cell. It contains all the vital information needed by the cell or the whole organism to function, grow and reproduce. This information is stored in the form of a genetic code on the chromosomes, which are situated inside the nucleus.

Proteins are the basic building materials of a cell, made by the cell itself. Looking at them in close-up they consist of a chain of amino-acids, small specific building blocks that easily link up. Though the basic structure of proteins is linear, they are usually folded and folded again into complex structures. Different proteins have different functions. They can be transport molecules (e.g. oxygen binding haemoglobin of the red blood cells); they can be antibodies, messengers, enzymes (e.g. digestion enzymes) or hormones (e.g. growth hormones or insulin). Another group is the structural proteins that form boundaries and provide movement, elasticity and the ability to contract. Muscle fibres, for example, are mainly made of proteins. Proteins are thus crucial in the formation of cells and in giving cells the capacity to function properly.

Chromosomes means "coloured bodies" (they can be seen under the light microscope, using a particular stain). They look like bundled up knots and loops of a long thin thread. Chromosomes are the storage place for all genetic-that is hereditary-information. This information is written along the thin thread, called DNA. "DNA" is an abbreviation for deoxyribo nucleic acid, a specific acidic material that can be found in the nucleus. The genetic information is written in the form of a code, almost like a music tape. To ensure the thread and the information are stable and safe, a twisted double stranded thread is used-the famous double helix. When a cell multiplies it will also copy all the DNA and pass it on to the daughter cell.

The totality of the genetic information of an organism is called genome. Cells of humans, for example, possess two sets of 23 different chromosomes, one set from the mother and the other from-the father. The DNA of each human cell corresponds to 2 meters of DNA if it is stretched out and it is thus crucial to organise the DNA in chromosomes, so as to avoid knots, tangles and breakages. The length of DNA contained in the human body is approximately 60,000,000,000 kilometres. This is equivalent to the distance to the moon and back 8000 times!

The information contained on the chromosomes in the DNA is written and coded in such a way that it can be understood by almost all living species on earth. It is thus termed the universal code of life. In this coding system, cells need only four symbols (called nucleotides) to spell out all the instructions of how to make any protein. Nucleotides are the units DNA is composed of and their individual names are commonly abbreviated to the letters A, C G and T These letters are arranged in 3-letter words which in turn code for a particular amino acid. The information for how any cell is structured or how it functions is all encoded in single and distinct genes. A Gene is a certain segment (length) of DNA with specific instructions for the production of commonly one specific protein. The coding sequence of a gene is, on average about 1000 letters long. Genes code for example for insulin, digestive enzymes, blood clotting proteins, or pigments.

How does a cell know when to produce which protein and how much of it? In front of each gene there is a stretch of DNA that contains the regulatory elements for that specific gene, most of which is known as the promoter. It functions like a "control tower," constantly holding a "flag" up for the gene it controls. Take insulin production (which we produce to enable the burning of the blood sugar). When a message arrives in the form of a molecule that says, "more insulin", the insulin control tower will signal the location of the insulin gene and say "over here". The message molecule will "dock in" and thus activate a "switch" to start the whole process of gene expression.

How does the information contained in the DNA get turned into a protein at the right time? Each gene consists of 3 main components: a "control tower" (promoter), an information block and a polyA signal element. If there is not enough of a specific protein present in the cell, a message will be sent into the nucleus to find the relevant gene. If the control tower recognises the message as valid it will open the "gate" to the information block. Immediately the information is copied-or transcribed-into a threadlike molecule, called RNA.

RNA is very similar to DNA, except it is single stranded. After the copy is made, a string of up to 200 "A"-type nucleotides-a polyA tail-is added to its end. This process is called poly-adenylation and is initiated by a polyA signal located towards the end of the gene. A polyA tail is thought to stabilise the RNA message against degradation for a limited time. Now the RNA copies of the gene leave the nucleus and get distributed within the cell to little work units that translate the information into proteins.

No cell will ever make use of all the information coded in its DNA. Cells divide the work up amongst one other-they specialise. Brain cells will not produce insulin, liver cells will not produce saliva, nor will skin cells start producing bone. If they did, our bodies could be chaos! The same is true for plants: root cells will not produce the green chlorophyll, nor will the leaves produce pollen or nectar. Furthermore, expression is age dependent: young shoots will not express any genes to do with fruit ripening, while old people will not usually start developing another set of teeth.

Gene regulation is very specific to the environment in which the cell finds itself and is also linked to the developmental stages of an organism. So if one want the leaves of poppy plants to produce the red colour of the flower petals he/she will not be able to do so by traditional breeding methods, despite the fact that leaf ells will have all the genetic information necessary. There is a block that prevents he leaves from going red. This block may be caused by two things:

— The "red" gene has been permanently shut down and bundled up thoroughly in all leaf cells. Thus the information cannot be accessed any more.

— The leaf cells do not need the colour red and thus do not request RNA copies of this information. Therefore no message molecule is docking at the "red" control tower to activate the gene.

Of course-you might have guessed-there is a trick to fool the plant and make it turn red against its own will. We can bring the red gene in like a Trojan horse, hidden behind the control tower of a different gene. But for this we need to cut the genes up and glue them together in a different form. This is where breeding ends and genetic engineering begins.

Breeding is the natural process of sexual reproduction within the same species. The hereditary information of both parents is combined and passed on to the offspring. In this process the same sections of DNA can be exchanged between the same chromosomes, but genes will always remain at their very own and precise position and order on the chromosomes. A gene will thus always be surrounded by the same DNA unless mutations or accidents occur. Species that are closely related might be able to interbreed, like a donkey and a horse, but their offspring will usually be infertile. This is a natural safety devise, preventing the mixing of genes that might not be compatible and to secure the survival of the species.

Fundamentals of Genetic Engineering

Genes are the substance within the nucleus of a cell that transmit genetic codes. In human beings and other bio-species, the synthesis of protein and the reproduction of cells and tissue take place according to the genetic blueprint contained in the genes. There are approximately 7,000 to 10,000 kinds of genes of differing sizes and properties in the human body. After several decades of research, people have gained a preliminary understanding of the structure and properties of about 1,500 small genes. However, knowledge of large genes, especially chromosomes, is very limited. The larger the gene, the more difficult it is to determine its structure and properties.

Chromosomes and other genes are composed of DNA. There are four types of nucleotides (DNA building blocks) labeled A, T, G, and C [according to the four types of nitrogenous bases occurring in the nucleotides, namely, adenine, thymine, guanine, and cytosine]. These four types of nucleotides combine in different sequences to form long chains. The sequencing of A, T, G, and C nucleotides is called the primary structure. The twisting of the long chains is called the secondary structure. Pairs of long chains couple together in a double helix structure. The G in one chain couples with the C in another chain, and the T in one chain couples with the A in another chain. A human chromosome can be as long as 3 billion nucleotides. The genetic codes are stored in these DNA sequences.

Genetic engineering aims to re-arrange the sequence of DNA in gene using artificial methods. To determine the DNA sequence on such a long chain, a high-speed, low-cost method is needed. Unfortunately, the currently available method (gel-electrophoresis) does not satisfy these requirements. It cannot provide sufficient information about the genetic structure. For example, the DNA chain in the gene of *E. coli* is approximately one million nucleotides long. Its DNA sequence analysis took 12 years. Thus, high-speed determination of DNA sequence in genes is critical to the development of genetics. Without such a technique, it will be impossible to understand the genetic mechanism, and GE will be built on thin ice.

Given this predicament, the scientific community, supported by the U.S. Congress has delegated substantial resources to be used in determining gene structure under the auspices of the Human Genome Project (HGP). In addition to improving the gel-electrophoresis method, HGP has also invested a huge amount of capital in supporting the discovery of a new method for DNA analysis. Several new mass-spectroscopic techniques, such as MALDI-TOF (matrix-assisted laser desorption and ionization-time-of-flight mass spectrometer) and FTMS

(Fourier transform mass spectrometer), have been developed and applied to analyze the DNA segment.

However, these methods run into difficulties when the DNA segment is larger than 200 nucleotides. As the size of the DNA increases, the sensitivity and resolution of the signal greatly decreases and analysis using these methods gradually becomes unreliable. To date this difficulty has not been overcome despite intensive research. What is more, the method of gel-electrophoresis has seen little improvement in recent years. Due to the overall lack of information about gene structure, our knowledge of the genetic mechanism is very limited.

With every new scientific discovery, there are always those who use it to seek profit. The field of genetics is no exception in this matter. Investors began getting into GE business several years ago. What is the current stage of development of GE? Roughly, it can be seen in three major fields.

1) *The determination of DNA sequence in chromosomes and other genes.* Work in this field is mainly supported by the U.S. government, which considers long-term development to be of national interest. Many private industrial enterprises are also extensively involved, since the patent owner of such an analytic technique will reap a huge return. Several years ago, there were many reports on developments in this field. The excitement has gradually subsided because scientists have realized the complexity of the problem. This problem is not going to be solved in the near future. There has been very little overall advance in this field during the past several years.

2) *Artificial horizontal gene transfer—a synthetic method of gene transfer between different species.* Since the structure and function of some small genes are relatively well known, biologists try to transfer these genes to other bio-species to improve their functions. Private enterprises have actively been testing this

method on animals and vegetables in order to obtain "super products." Government-supported research institutes mainly use horizontal gene transfer to obtain knowledge about the genetic mechanism. Since the genetic mechanism is a very complicated system, they can mostly conduct blind tests by means of horizontal gene transfer. There are many unknown factors in this field, regardless of whether the method used is direct insertion of genes or simple mixing of genes. The probability of success is very small, and only a few products exist. The possible side effects of these GE experiments are still unclear. There has not been much successful advance in this field.

3) *Cloning.* 'Dolly'-sheep is genetically duplicated using a complete set of chromosomes from an adult sheep. This success was based on a large number of failures. But scientists have not been able to repeat it. The validity of scientific results is based upon their reproducibility. Since the experiment has not been repeated, many people doubt its validity. About 50% of the scientific community is not convinced by the result. It would be extremely difficult to clone a human being even in the absence of pressure from social objections.

Some scientists believe that, since genetic codes determine the appearance, personality, health, and aging process of human beings, if that genetic information in the chromosomes could be decoded and the genetic mechanism were understood, we could potentially control and improve our health, quality of life, and the biochemical processes in our bodies. In other words, we could control our own fate. Also, we'd be able to improve the genes of other animals and vegetables so that they could serve humankind better. At first sight, these ideas seem reasonable and attractive. However, careful analysis reveals that they are based upon an incorrect theory—the theory of gene determinism.

Gene determinism says that gene structure determines everything about life. This theory is based on an incomplete view of things. It not only denies the mutual interaction between the body and mind, but also neglects the fact that genes survive through mutations. Buddhists know that body and mind are inseparable. They interact and influence each other. The structure and function of chromosomes and genes can change as the mind changes. It would be erroneous to reduce life to its material aspects. A person's traits and health depend on his/her state of mind and environment as well as on gene structure. Thus the theory of gene determinism is merely a form of materialism and does not reflect reality. Not only a person's thoughts and actions in the present life, but also his past life history exerts a great influence on his metabolic processes, and his genetic inheritance is only of secondary importance.

Although the chromosomes and genes are the blueprint of protein synthesis, not all parts of such a long chain are used in the synthesis of protein. Generally, only a small segment of the chain is essential. In different organs or under different conditions, different segments are used in synthesis. Protein synthesis and cell reproduction are extremely complicated processes that are determined not by a single DNA chain, but by various other conditions working together. Human mental factors, such as emotions and habits, play an important role in these processes. Sometimes they become the decisive factors. Therefore, it is impossible for GE to fulfill its promise of improved health and quality of life for humankind. A large number of examples in psychology can be used to prove this point.

Another erroneous assumption of gene determinism is that the structure of genes rarely or never changes. Scientific research in recent years indicates that the structure of genes is dynamic and subject to change during interaction with the environment. These changes can also be passed on to the next generation. If the gene structure is dynamic, then the design of gene structure becomes insignificant. Given the correlation between gene function and psychological factors as well as the influence of the environment on gene structure, we may conclude that attempts to design or alter gene structure will not achieve the "expected goal."

By the early 1960s geneticists had a good basic picture of genetics. However, they did not have any way to obtain and study the genes themselves except through the laborious procedures of mating and testing mutants. These procedures take an immense amount of time, money and effort and yield disproportionately meager results. Further, there were no techniques to examine the genes at a molecular level, yet unless this was done the ultimate questions could never be answered. Because of these problems older geneticists were leaving the field and younger, ambitious ones, were looking elsewhere for their creative outlet. Then, as so often happens in science, a series of totally unrelated serendipitous discoveries were made which propelled the genetic revolution upon the world. It all started with a seemingly odd result obtained with some phage.

In the 1950s it had been noted that if one grew a particular bacteriophage on a particular bacterial mutant host strain a and then infected a bacterial mutant-strain b, of the same species, with the phage a, the yield of phage from strain b was very low. However, if you took the few phage b that were produced and infected strain b with them, the phage yield was now normal. Subsequently, in the 1960s, it was found that the phage DNA that grew on strain b had been chemically modified so that it could not be cleaved by a DNAase in strain b. The DNAase involved in this cleavage was found to be somewhat specific, in that it mainly cut the DNA at certain sequences; unless these sequences had been chemically modified by enzymes in the cell. All previous

DNAases cut DNA randomly, so this finding suggested that DNA could be cut up into specific fragments which would always contain the same set of genes between the cut-sites. However, these first "specific" DNAases did not prove as specific as hoped. Finally, in 1970 hamilton smith accidentally found that a DNAase from the bacterium, *Haemophilus influenzae*, cut DNA only at unique DNA sequences known as palindrome sites.

Palindromes

In DNA a palindrome site is a sequence of base pairs in double stranded DNA that reads the same backwards and forward across the double strand. For example, the sequence of base pairs GAATTC is a palindrome because both sequences of the double strand read the same when read from either their respective "G" or "C" ends (complementary strand = CTTAAG). The enzymes that cut these specific sites are called restriction enzymes. Based on the types of cuts they make, there are two types of restriction enzyme: One group cuts straight across the double stranded DNA, producing blunt ended DNA (Fig. 1).

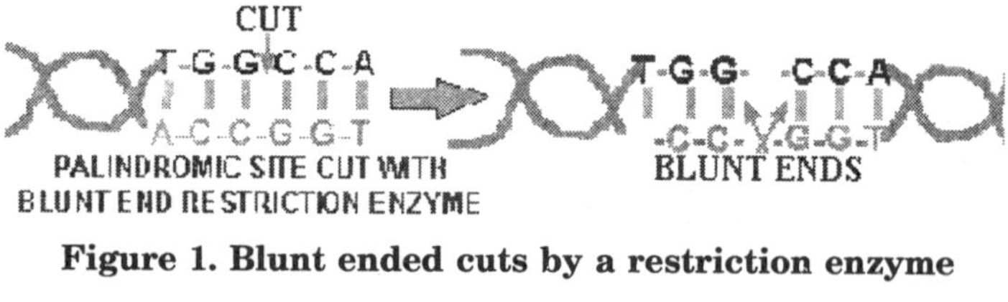

Figure 1. Blunt ended cuts by a restriction enzyme

Another type cuts the strand of DNA off the center of the palindrome sites, but between the same two bases on the opposite strands. This leaves one or more bases overhanging on each strand and such ends are called sticky ends (Fig. 2) because they form hydrogen bonds with their complementary cut counterparts. For example, in the base sequence GAATTC, the restriction enzyme that cuts this site cleaves the DNA between the "G" and the "A" on each complementary strand, thus leaving the overhangs of AATT & TTAA respectively.

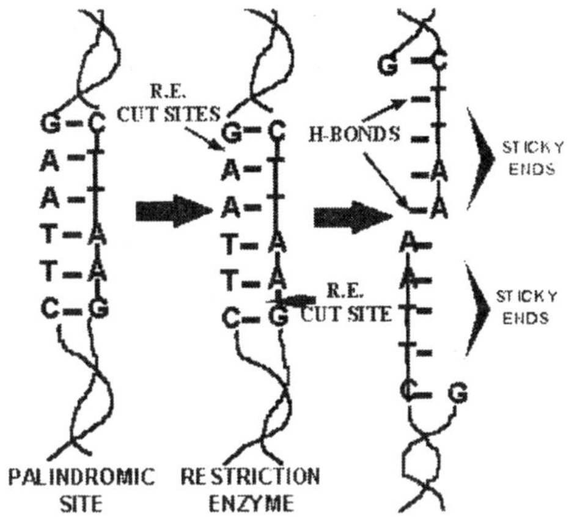

Figure 2. Sticky ended cut with restriction enzyme

Over 200 different restriction enzyme have been isolated that cut at many different specific DNA sequences. They are all quite specific and this specificity is the key to their use. Restriction enzymes that cut at 4, 5, 6, 8, 9, and 11 base pair palindrome sites have been found. It follows that the more base pairs in the palindrome site the less likely a site is to exist statistically. That is, a 4-base-cutting restriction enzyme cuts the same DNA many more times than an 8-base-cutter does. The odd numbered cutting sites are not true DNA palindromes, in that the central base pair reads different in the two directions.

The final important component in genetic engineering is an enzyme called ligase. Ligase is an enzyme that covalently joins the sugar-phosphate backbone of bases together. Ligase is also involved in DNA replication. In effect ligase reverses the action of restriction enzymes, which break the sugar-phosphate backbone-bonds. Ligase requires energy to form these bonds (Fig. 3).

Ligase will join either "sticky" ends or "blunt" ends, but it is more efficient at closing sticky ends because the "sticky overhangs" of these ends adds stability which holds the strands together while the ligase works, whereas the blunt ends are less stable. The sticky ends must have the same "base overhang", but any blunt end can be joined to any other blunt end.

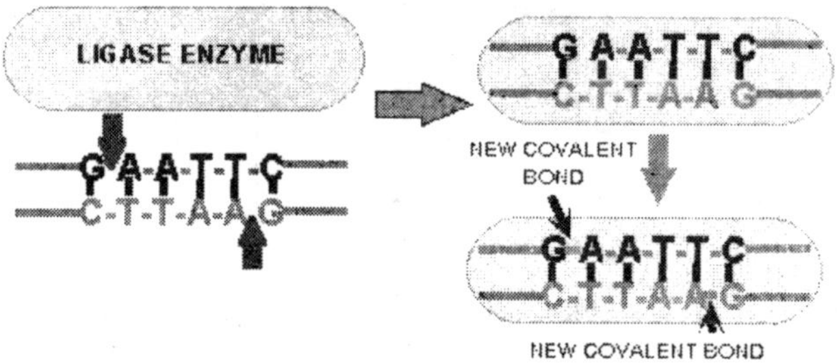

Figure 3. The ligase reaction. The sticky ends of two DNA strands line up according to their base pair hydrogen bonding. The ligase binds to the hydrogen-bonded DNA strands and forms covalent bonds between the two DNA ends, thus joining them. Similarly, if the blunt ends of DNA come together the ligase can form covalent bonds between them.

Steps in Genetic Engineering

Basically genetic engineering includes the following the components:

1) Double-stranded DNA contain palindrome restriction enzyme sites.

2) Many types of restriction enzymes cut or cleave palindrome sites in the double-stranded DNA sample forming either a sticky or a blunt end.

3) The enzyme ligase plus an energy source fuses or joins two sticky or two blunt ends of DNA together

In genetic engineering the idea is to move a "gene of interest" from its original or Source DNA molecule into another or Target DNA molecule for some specific purpose. Those of you who are computer aficionados will note the similarity in terminology in describing the moving of computer files.

The process of moving a gene from one DNA molecule to another DNA molecule is called Cloning. In the cloning process a fragment of DNA, containing a gene-or sequence-of-interest, is cloned (moved) into a vector where it can subsequently be grown (copied) in large quantities and manipulated in a variety of ways. The most common vectors are bacterial plasmids, but viruses and self-replicating units in eukaryotic cells are also employed as vectors.

The process of cloning is straight forward and only requires the following four steps and can be done in a couple of hours or less.

— Isolating the source and vector DNA. The DNA's must be relatively free of contaminating materials which interfere with the subsequent enzymatic steps.

— Both the source and vector DNA are cut with Restriction Enzymes. When sticky ends are formed the DNA is cut with the same restriction enzyme(s), but restriction enzyme that produce blunt ends also work well in cloning.

— The vector and source DNA are mixed with a Ligase System and covalently bonded together.

— Finally, the Ligated DNA is Transformed into a host cell. Usually the host cell is a Competent bacterium, but increasingly eukaryotic cells are being used. After suitable growth has occurred the host cells are examined for the presence of the source or cloned DNA in its cytoplasm.

The entire cloning process often takes less than a day and is carried out using volumes of 1,000 microliters or less. The examination of the transformed host for the gene-of-interest may take an additional day or so. A gene which has been successfully transferred in this way is said to have been cloned. The process is referred to as "cloning", "recombinant DNA technology", or "recombinant DNA". These steps are summarized in the figure 4.

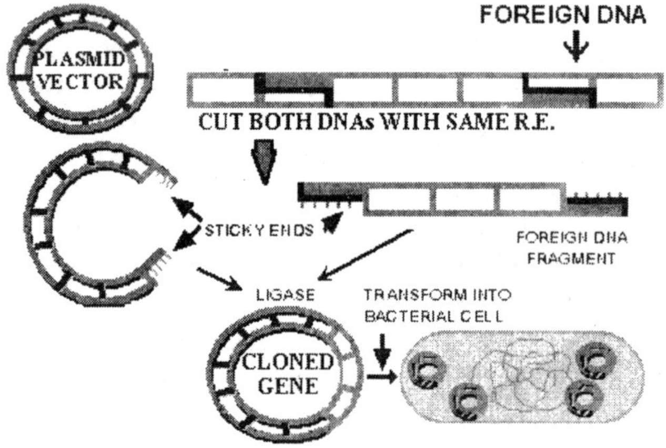

Figure 4. Cloning

Uses of Cloned DNA

To do molecular biology on a single gene it is necessary to isolate it from all the other genes. This is not always easy considering that an average bacterium contains 1,000 to 4,000 genes, while eukaryotic cells may contain >100,000 genes. A cloned gene can be amplified by making many copies of the clone so as to produce large enough quantities to study it in detail. The usual things that are done to an amplified gene are:

— *sequencing*: sequencing determines the base pair sequence of a gene. By reading the 3-letter code, sequencing also describes the amino acid sequence translated from that gene.

— *mutation*: the gene base pair sequence can be changed in specific ways and the modified gene can be inserted back into its original host to see what each specific mutation does. Whereas, spontaneous mutation is random, techniques are now available that make it possible to change any codon within a gene to any other codon. Therefore, it is possible to study the effects of single amino acid changes on the function of the gene product, which is, after all, the ultimate purpose of the exercise.

— to replace a mutated form of the gene in the original host cell with a healthy form of the gene to see exactly what it does in its intended place of residence.

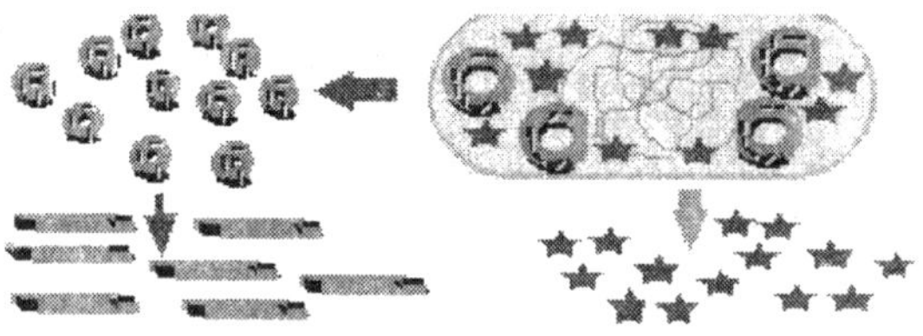

Figure 5. Amplification of Cloned DNA or its Product

— to use the amplified gene to make humongous quantities of the gene product for commercial purposes (Fig. 5). This is how products like human insulin, human growth hormone, plasminogen activator, interlukins and the bovine milk hormone are all produced.

— to insert the gene into another species for some purpose. Such animals or plants are said to be transgenic. For example, many crops are protected against caterpillar larvae because a gene from a bacterium has been inserted into their cells. This gene produces a protein that is highly, and specifically, toxic to the larvae of many moths and butterflies that attack food crops.

Transgenic Organisms

Transgenic means "across species". The above techniques give us the ability to move viral genes into humans, human genes into plants, plant genes into viruses etc. etc. there is a flood of transgenic plants in the approval pipeline; two types of transgenic tomatoes are in our local supermarkets, transgenic squash are coming soon and a number of transgenic crops are currently protected from insect larvae by the protein described above.

There is a down side of genetic engineering. The same procedures that can be used to produce a better crop, or human insulin, can also be employed to make a more powerful pathogen. There is evidence that some countries have or are considering doing this. For example, Iraq was apparently growing bacterial pathogens for biological warfare, but it was deterred from using them when we threatened atomic retaliation. It is only a small jump in "reasoning" to consider engineering a "better pathogen" with which to destroy a hated enemy.

DNA Fingerprinting

The three basic principles required to understand DNA fingerprinting are:

— base-pairing of AT (AU) & GC is the basic principle of this procedure. If you understand this all else fall easily into place.

— specificity of enzyme activity is the second crucial principle to understanding DNA fingerprinting. This refers to the cutting of DNA by specific restriction enzymes at unique palindrome sequences.

— the recognition that a change in a single base pair (a mutation) can either make a restriction enzyme-site where one did not exist previously or it can remove or eliminate a restriction enzyme site from a gene. An analogy would be to "mutate" your phone number by one letter; callers would get a different person.

To understand DNA fingerprinting the double stranded DNA molecule must be viewed as a chain with random restriction enzymes-sites (palindromes) located along its length. If each of these various unique restriction sites were marked with a different color, the DNA would appear as a random multicolored stripped ribbon. If one set of unique restriction enzyme sites were colored dark and the appropriate dark-restriction enzyme added, the DNA would be cleaved into a series of various sized fragments depending on where the red-sites were located along the DNA molecule (Fig. 6).

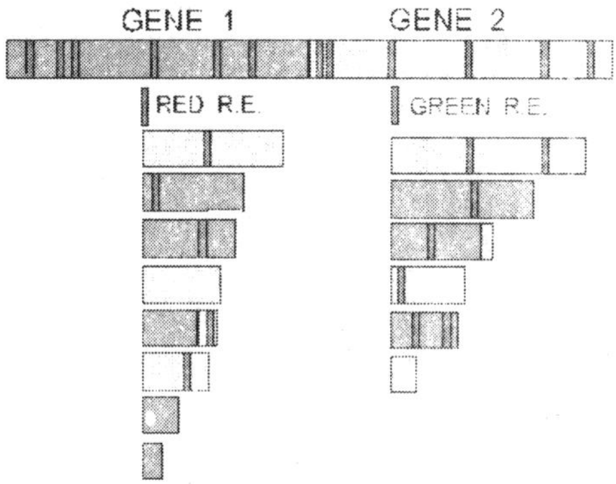

Figure 6. DNA Fingerprinting. In the figure there are two genes with the palindrome sites for two different restriction enzymes marked as grey or dark bars. The number and size of the DNA fragments that would result if the DNA containing these two genes were cut with the respective restriction enzymes are shown.

It is easy to understand that the group of different sized fragments would be unique for two DNA strands with the red-restriction sites located at different places along their respective DNA double strands. The addition of a green-restriction enzyme that cuts green-sites would produce a different group of sized fragments. Each of these groups of restriction enzyme-produced fragments would represent a unique fingerprint of that DNA.

The DNA fragments are visualized using the technique known as gel electrophoresis. Briefly, porous gels composed either of a plastic called acrylamide or of a derivative of agar, called agarose, are used to separate DNA fragments based on size. The gels are prepared with wells into which the DNA fragments are added. The gels are submerged, under an electrolyte buffer solution between a positive and a negative electrode; the DNA-containing solutions are added to the wells and the current is turned on. The DNA fragments are negatively charged so the wells containing them are placed closest to the negative electrode. When the current is turned on the DNA moves through the pores in the gel towards the positive electrode. The shorter fragments move fastest because they are able to navigate through the pores of the gel more easily, whereas the longer DNA fragments move proportionally more slowly through the pores. The result is a separation of the DNA fragments based on their length (size). The DNA are stained by dyes that binds to the DNA and fluoresces strongly when exposed to UV-light. The various sized DNA fragments appears as a series of bands that resemble a bar code.

DNA fingerprinting is now good enough to identify individuals with the same accuracy as classical fingerprinting because of the improvements of the procedures.

Mutations and DNA Fingerprinting

DNA fingerprinting works because of mutations which change the genetic code, randomly making and eliminating restriction enzyme sites randomly throughout the DNA. Because of these variation in genes between individuals the likelihood that the restriction enzyme fragment pattern of two individuals being the same is very very small unless they are identical twins. To determine a DNA fingerprint a small amount of DNA-containing material from an organism is required. This could be obtained from any tiny quantity of tissue or blood, even the material from a single hair follicle is sufficient. The variation in the gel fragments from different DNA from members of the same species is called restriction fragment length polymorphism (RFLP) analysis.

In Figure 7 the gel pattern of a hypothetical analysis of O.J.'s DNA vs. a sample of unknown DNA is depicted. The two isolated DNA were treated with the same restriction enzyme and the fragments separated on an agarose gel. Two unique restriction enzyme fragment patterns are seen. The conclusion is that the blood samples came from different individuals. The source of the DNA clearly influences the significance of these results. For example, if the two samples were collected at the scene of the crime it would mean one thing, but if the unknown DNA sample came from O.J.'s shirt and it matched that of one of the victims at the scene of the crime, the data takes on more significance.

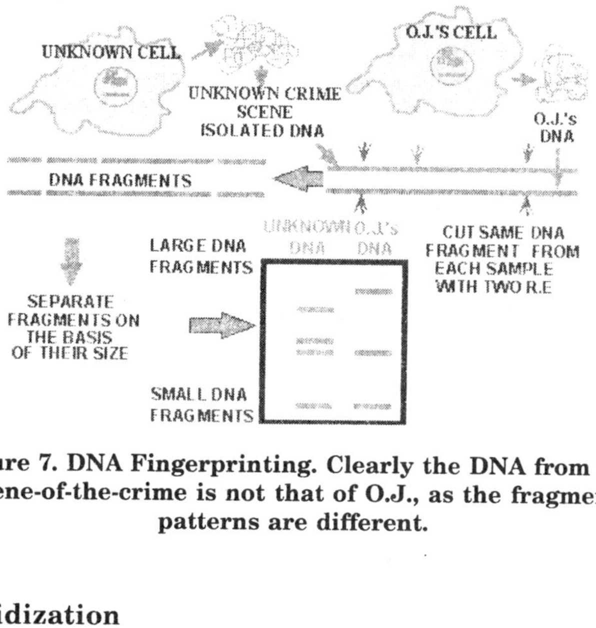

Figure 7. DNA Fingerprinting. Clearly the DNA from the scene-of-the-crime is not that of O.J., as the fragment patterns are different.

Hybridization

When genomic DNA is treated with a restriction enzyme and separated on a gel, what is seen is a continuous smear composed of 1,000s of individual DNA fragments. This problem is solved by a technique known as hybridization. Hybridization depends on the basic principle of hydrogen bonding between GC & AT bonds in nucleic acids. The principles in the hybridization steps are:

— DNA strands are separated by breaking the hydrogen bonds between the bases with heat to produce single strands.

— Strand separation is strictly dependent upon two factors: the temperature and the number of hydrogen bonds.

— When the temperature drops, single DNA strands in the same solution that are complementary spontaneously come together by pairing up through their respective AT & GC associations. The strength of their subsequent association depends on the temperature and the total number of matching base pairs.

— The test or sample DNA is usually fixed or bound to a solid surface in a single-stranded state. A piece of DNA of known sequence to which something is attached that can be detected or seen is then mixed with the bound DNA and the two are incubated together under conditions that will allow complementary DNA sequences to join through the appropriate hydrogen bonding. The DNA molecule which can be detected is called a probe.

That is, the more hydrogen bonds there are the higher the temperature must be to completely separate two DNA strands and to keep them separated. For example, if there were two stands of DNA composed of 200 perfectly complemented base pairs, then it would take a temperature of 58°C. to separate them. However, if only 199 of the base pairs were correct, it would only take 57°c. to separate them; if there were only 100 correct base pairs the DNA strands would fall apart at ~29°c; i.e., fewer bonds require less heat (energy = lower temperatures) to rupture them (Fig. 8).

Hybridization is carried out as illustrated in Fig. 9:

1. the isolated DNA is fragmented (cut) with restriction enzymes.
2. the cut-DNA is separated into fragments according to size on a gel.
3. the DNA fragments are transferred, in position, to a membrane to which they tightly bind.
4. the membrane-bound-DNA is soaked in a solution containing probe-DNA that is labeled with something that can be detected.
5. the membrane-bound-DNA-probe mixture is incubated at a temperature designed so that only the probe DNA strands that have a high degree of complementarity (base pair matching) to DNA strands bound to the membrane will be able to bind (hydrogen-bond) to each other.
6. the unbound probe is washed away and the position of the bound probe is determined using its detection system of the probe.

The entire procedure is called hybridization. Hybridization works because of the sequence of the probe. A given probe will only bind with DNA that is attached to the membrane if it finds a matching or complementary DNA base pair sequence that forms enough bonds to remain together under the heating conditions used. Thus if no matching complementary sequences are found, no probe will bind. If one or two of the 1,000s of the bound DNA fragments contain a sequence complementary to the sequence in the probe, the probe will bind only to these fragments and thus become attached to the membrane through this association with the complementary membrane-bound DNA.

In theory DNA fingerprinting will work with much smaller amounts of material than a classical fingerprint &

DNA lasts much longer than classical fingerprints. DNA-containing samples that are many years old are still usable.

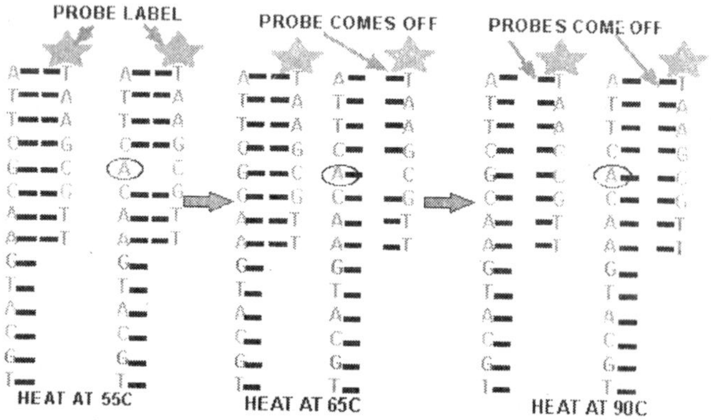

Figure 8. Effect of Temperature on Hybridization. In this figure a short piece of DNA, called the Probe (dark star) is mixed with two different long DNA strands. One of these long strands contains a sequence of base pair that exactly matches that of the probe, whereas the other differs with respect to a single base pair (oval round). At 55°C the probe binds to both long DNA molecules, however at 65°C the probe does not bind to the long DNA molecule with a single base pair mismatch. The probe is unable to bind to either long DNA molecule at 90°C.

Only very tiny quantities of DNA are required in order to carry out a highly accurate test. For example, dried blood, semen, spit, skin etc. on samples stored in dusty files for years are still usable. Samples of mixed DNA's can also be used. DNA containing evidence is much harder to clean up at a crime scene than other evidence, like classical fingerprints.

Even if there is a perfect match between DNA, you can not say how the DNA containing sample got there or when. In the O.J. trial a valid question was raised about the possibility of evidence being planted. What makes this charge so powerful is the extreme sensitivity of the procedure. That is, it would not be difficult for someone to collect a small quantity of blood, body fluid or tissue from a victim and place this material where it would incriminate an innocent person.

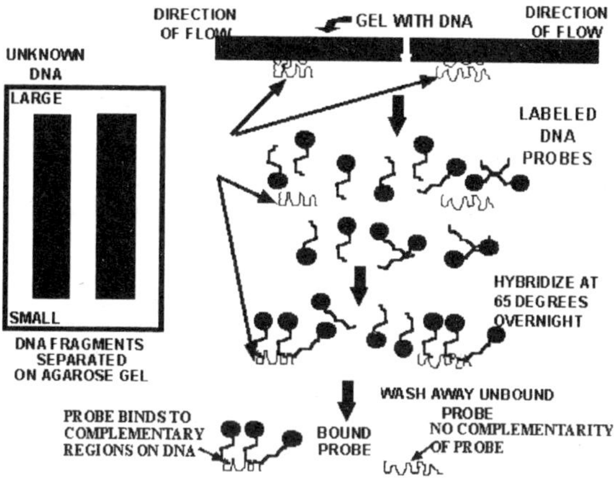

Figure 9. The hybridization procedure. The DNA samples are cut with R.E. and the fragments separated on a gel. The gel is placed on top of a DNA-binding protein and the DNA-fragments are washed directly onto the membrane where they stick. A short piece of DNA (the probe), to which a reporter molecule is attached, is heated with the bound DNA overnight. During this time the probe will bind to any complementary DNA attached to the membrane. All the unbound probe is subsequently washed away. The reporter is visualized so the location of the sample DNA fragments complementary to the probe DNA can be seen.

There are documented cases where quantities of drugs, marked money etc. have been "planted" and innocent people convicted of crimes because of these actions. Finally, blood that is mixed with the wrong chemicals or is degraded is difficult to analyze accurately (Fig. 10).

Polymerase Chain Reaction (PCR)

The polymerase chain reaction (PCR) won its discoverer a Nobel Prize in 1993. Kary Mullis has recorded the thought process he went through as he conceived of the idea and he states that he knew immediately that he was going to become very "famous" because of this discovery. Everyone else assumed that he would win the Nobel Prize as soon as they heard of the PCR reaction. The ironic thing about this discovery, is that all the scientific information had been around for approximately 20 years and had been overlooked by the greatest scientific minds in the world. This discovery is now seen as such an obvious concept that some scientists question if Mullis deserved the Nobel Prize for such a "simple" and "obvious" discovery.

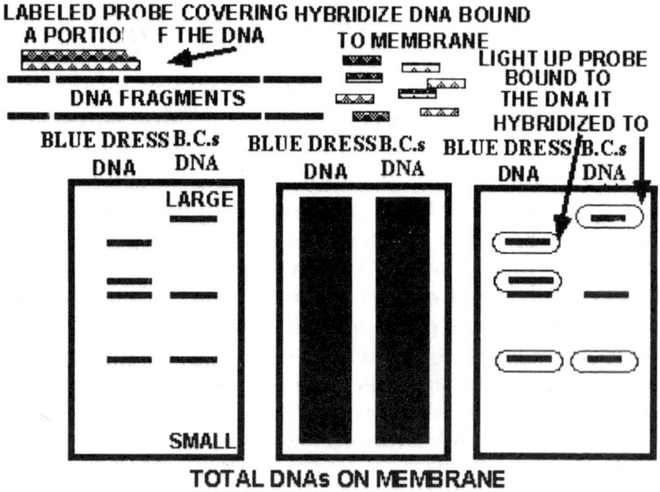

Figure 10. Results of probing the separated fragments of two DNA samples. The original gel, shown in the middle, produces a smear of genomic DNA fragments. Within those 1000's of fragments are the fragments shown on the left gel cut from a section of the two original DNA molecules shown in the upper left. When the smear of fragments, fixed to a membrane, are hybridized with the probe, the probe will hybridize only to those fragments containing sequences of DNA that complement sequences present in the probe.

The PCR is another one of these discoveries, like that of the structure of DNA, the transforming factor determining the chemical nature of DNA, and restriction enzymes, that generated a quantum leap in science. Almost the instant after PCR is explained to a biological scientist, ideas as to its uses begin to pour from all but the most dull scientific mind. Even today, years after its discovery, we are still developing new uses for PCR.

The irony of the PCR is that living organisms have been doing it since they evolved in the primeval organic soup of this planet 3.5 billion years ago and scientists have been aware of the general details of this process for approximately 50 years. To put it succinctly, the PCR does in the test tube what every bacterium does in its tube of media or on an agar-plate and each of us do every day; we all produce billions of exact copies of our own DNA; amplifying our DNA millions of time. The enzyme DNA polymerase was discovered in the 1950s and our knowledge of the process has been increasing ever since. This means that thousands of scientists have studied DNA replication for 40 years without tumbling to PCR.

The basic principle of PCR is shown in Fig. 11. It has been know for a long time that DNA polymerase requires a short strand of DNA or RNA called a primer to "prime" the start of DNA replication. Mullis's genius was that he reasoned "That if you added the following components to a test tube containing a single DNA molecule, you could replicate & amplify that DNA molecule many million fold in a short time.

The components are:

— A sample of the target DNA to be copied. In theory only a single molecule is needed.

— A set of short (15 to 40 bases) single stranded primers of DNA, in excess, that will bind to complementary regions of the opposing stands of the target DNA molecule. These primers bracket the region of DNA to be amplified.

— An excess of the 4 nucleotide triphosphates, ATP, GTP, CTP, TTP.

— The enzyme, DNA polymerase.

— Various buffers and cofactors like magnesium ions required by DNA polymerase.

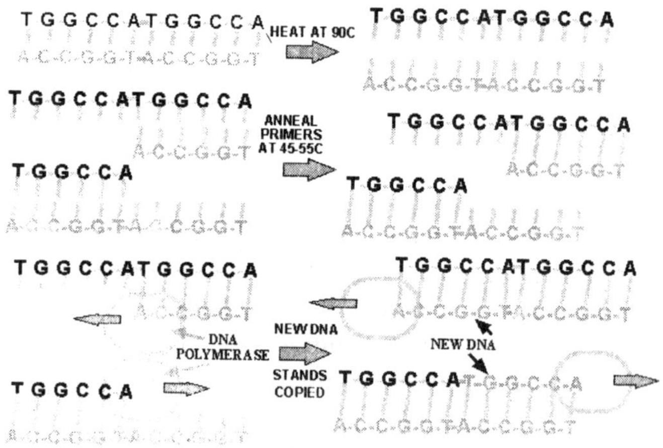

Figure 11. The polymerase chain reaction

The final trick was to get the two target DNA strands apart (separated) so the primers could bind and the DNA polymerase could do its thing. It had been known for ~50 years that heat separates DNA strands and that complementary strands then rejoin through base pairing when the temperature is subsequently lowered. So Mullis heated his mixture of target DNA, primers and triphosphate nucleotides to about 90°C for a few minutes to separate the target DNA. He then lowered the temperature enough to allow the primers, which were small and in vast excess, to bind (anneal) to their respective complementary target DNA base pair-sequences. At this point he added DNA polymerase and allowed the polymerization reaction with the triphosphate nucleotides to occur. That is, the DNA polymerase filled in the missing portion of each strand making two new double stranded regions of DNA.

Each entire PCR cycle takes only 2 to 10 minutes. However, there was one problem with the system devised by K. Mullis, which was that the DNA polymerase he used was destroyed by the heating process. Mullis had to add new DNA polymerase for each round, which was time-consuming and expensive. This problem was solved by using heat-resistant DNA polymerase that only needed to be added once. Thermophilic bacteria that live in boiling hot springs. This is another case of serendipitous basic science turning out to be important.

The study of thermophiles might seem to be a waste of money and time to many people as these bacteria, while certainly interesting, don't cause disease in man or any other life form. The argument could be made that a scientist could better spend his/her time working on cancer or some other terrible disease that afflicts humankind. However, it turns out that since thermophilic DNA polymerases tolerate high temperatures (e.g. 90°C) for long periods without being destroyed, they are the perfect solution to the PCR DNA polymerase problem. Thus today PCR has become a revolutionary tool because of scientists who studied these odd thermophiles.

The standard PCR reaction is run through about 30 cycles in a couple of hours which results in the amplification of the original DNA by over a 109 fold. Thus a single specific DNA region can be amplified to yield sufficient quantities to do anything that can be done with bulk-isolated DNA. In the case of crime evidence this means that the DNA in a single hair follicle, a single drop of semen or blood is sufficient to prove that an individual was present at the scene of a crime. Indeed any piece of evidence that contains one or more moderately long DNA fragments can be amplified with PCR and its RFLP determined for identification purposes or maybe to build a dinosaur eventually. However, at present the maximum limit for amplification is approximately 42 kilobases (42,000).

Applications of Genetic Engineering

One of the best known applications of genetic engineering is that of the creation of genetically modified organisms (GMOs). There are potentially momentous biotechnology applications of GM, for example oral vaccines produced naturally in fruit, at very low cost. This represents, however, a spread of genetic modification to medical purposes and opens an ethical door to other uses of the technology to directly modify human genomes.

These effects are often not traceable back to direct causes in the genome, but rather in the environment or interaction of proteins. The means by which 'genes' (in fact DNA strands that are assumed to have discrete effects) are detected and inserted are inexact, including such means as coating gold particles with DNA to be inserted and literally firing it at strands of target DNA, which is guaranteed to cause insertions in at least some random locations, which can on rare occasion cause unplanned characteristics.

Similar objections apply to protein engineering and molecular engineering for use as drugs. However, a single protein or a molecule is easier to examine for 'quality control' than a complete genome, and there are more limited claims made for the reliability of proteins and molecules, than for the genomes of whole organisms. While protein and molecule engineers often times acknowledge the requirement to test their products in a wide variety of environments to determine if they pose dangers to life, the position of many genetic engineers is that they do not need to do so, since the outputs of their work are 'substantially the same as' the original organism which was produced by the original genome(s).

An extreme ambition of some groups is human enhancement via genetics, eventually by artificial intelligence or molecular engineering.

Genetic sequencing which is used to identify each base in DNA is exceedingly cheap. As of mid-2005, it cost 1/10 of 1 cent to sequence a single base. At the current rate of price decrease, the entire human genome could have been sequenced for less than 100 U.S. dollars.

Genetic Engineering and Research

Although there has been a tremendous revolution in the biological sciences in the past twenty years, there is still a great deal that remains to be discovered. The completion of the sequencing of the human genome, as well as the genomes of most agriculturally and scientifically important plants and animals, have increased the possibilities of genetic research immeasurably.

Expedient and inexpensive access to comprehensive genetic data has become a reality, with billions of sequenced nucleotides already online and annotated. Now that the rapid sequencing of arbitrarily large genomes has become a simple, if not trivial affair, a much greater challenge will be elucidating function of the extraordinarily complex web of interacting proteins, dubbed the proteome, that constitutes and powers all living things. Genetic engineering has become the gold standard in protein research, and major research process has been made using a wide variety of techniques, including

— loss of function, such as in a knockout experiment, in which an organism is engineered to lack one or more genes. This allows the experimenter to analyze the defects caused by this mutation, and can be considerably useful in unearthing the function of a gene. It is used especially frequently in developmental biology. A knockout experiment involves the creation and manipulation of a DNA construct *in vitro*, which, in a simple knockout, consists of a copy of the desired gene which has been slightly altered such as to cripple its function. The construct is then taken up by embryonic stem cells, where the engineered copy of the gene replaces the organism's own gene. These stem cells are injected into blastocysts, which are implanted into surrogate mothers. Another method, useful in organisms such as drosophila, is to induce mutations in a large population and then screen the progeny for the desired mutation. A similar process can be used in both plants and prokaryotes.

— gain of function experiments, the logical counterpart of knockouts. These are sometimes performed in conjunction with knockout experiments to more finely establish the function of the desired gene. The process is much the same as that in knockout engineering, except that the construct is designed to increase the function of the gene, usually by providing extra copies of the gene or attracting more frequent transcription.

— 'tracking' experiments, which seek to gain information about the localization and interaction of the desired protein. One way to do this is to replace the wild-type gene with a 'fusion' gene, which is a juxtaposition of the wild-type gene with a reporting element such as green fluorescent protein that will allow easy visualization of the products of the genetic modification. While this is a useful technique, the manipulation can destroy the function of the gene, creating secondary effects and possibly calling into question the results of the experiment. More sophisticated techniques are now in development that can track protein products without mitigating their function, such as the addition of small sequences which will serve as binding motifs to monoclonal antibodies.

Applications in Health Sector

Signal transduction Gene Therapy (STGT)

Signal transduction refers to the most common method of regulation of complex cell activity. It involves an external signal molecule (ligand) being bound by a specific cell's membrane receptor. The receptor, in turn triggers changes in the internal command system of the cell causing it to undertake a new activity so long as the ligand is present. A powerful new therapy using STGT is being developed whereby part of the genome of a common cold virus is replaced with:

1. A gene or genes that make a therapeutic chemical (e.g. hormone, cytokine etc.) and
2. A gene or genes that regulates the production of the therapeutic genes and that is itself "switched on" by a chemical that can be given to the patient.

In this procedure the patient is infected by the modified virus. The modified viral DNA then remains in the patient's cells. The therapy gene remain inactive until the "triggering" substance was given externally. The amount of triggering substance would determine the degree to which the therapy gene was turned on (like a dimmer switch).

Blocking the Damage of Arthritis

A cytokine known as the tumor necrosis factor (TNF) is known to be responsible for much of the effects of rheumatoid arthritis. It induces immune cells to produce substances that attack the host's own body cells causing the painful inflammation that accounts for most of the problems of rheumatoid arthritis. The TNF acts by binding to a specific membrane-bound protein receptor on the immune cells. The receptor protein, which extends into the cell cytoplasm undergoes a change in structure which causes it to react with other proteins within the cytoplasm which sets up a signal cascade that ends in the translation and transcription of the genes producing and releasing the damaging material.

Using GE technology a truncated (shortened) form of the receptor gene has been cloned. This gene makes a soluble form of the receptor that contains the TNF binding site. When large quantities of the material is injected into an arthritis patient it binds the TNF thus preventing it from binding to the cellular-associated receptor thus preventing (or minimizing) the signal from being sent.

Ethical and Social Issues of Genetic Engineering

Proponents of genetic engineering argue that the technology is safe, and that it is necessary in order to maintain food production that will continue to match population growth. However, others argue that food distribution, not production, is the biggest problem, citing that the population growth is actually a result of uneven distribution of food (and wealth).

Others oppose genetic engineering on the grounds that genetic modifications may have unforeseen consequences, both in the initially modified organisms, and their environments. For example, certain strains of maize have been developed that are toxic to plant eating insects. However, when those strains cross-pollinated with other varieties of wild and domestic maize, the relevant genes were passed on. This introduced a new gene into the gene pool of the maize population outside of the crop field. The ecological and environmental effects of transgenic plants are constantly being investigated.

Anti-genetic-engineering activists say that with current recombinant technology there is no way to ensure that genetically modified organisms will remain under control, and the use of this technology outside of secure laboratory environments carries unacceptable risks for the future.

Some fear that certain types of genetically engineered crops will further reduce biodiversity in the cropland; herbicide-tolerant crops will for example be treated with the relevant herbicide to the extent that there are no wild plants ('weeds') able to survive, and plants toxic to insects will mean insect-free crops. This could result in declines in other wildlife (e.g. birds) which depend on weed seeds and/or insects for food resources. The recent farm scale studies in the UK found this to be the case with GM sugar beet and GM rapeseed, but not with GM maize.

Proponents of current genetic techniques as applied to food plants cite the benefits that the technology can have, for example, in the harsh agricultural conditions of third world countries. They say that with modifications, existing crops would be able to thrive under the relatively hostile conditions providing much needed food to their people. Proponents also like to cite golden rice, a genetically engineered rice variety that contains elevated vitamin A levels. There is hope that this rice may alleviate vitamin A deficiency that contributes to the death of millions annually.

Proponents say that genetically engineered crops are not significantly different from those modified by nature or humans in the past, and by extension are as safe or even safer than such methods. There is gene transfer between unicellular eukaryotes and prokaryotes. There have been no known genetic catastrophes as a result of this. They argue that animal husbandry and crop breeding are also forms of genetic engineering that use artificial selection instead of modern genetic modification techniques. It is politics, they argue, not economics or science, that causes their work to be closely investigated, and for different standards to apply to it than those applied to other forms of agricultural technology.

Many opponents of current genetic engineering believe the increasing use of GM in major crops has caused a power shift in agriculture towards Biotechnology companies gaining excessive control over the production chain of crops and food, and over the farmers that use their products, as well. Many proponents of current genetic engineering techniques believe it will lower pesticide usage and has brought higher yields and profitability to many farmers, including those in third world countries. A few GM licenses allow third world farmers to save seeds for next year's planting.

References

1. *Adelberg EA (1960) Papers on Bacterial Genetics*. Boston: Little, Brown and Company.
2. Blatt RJR (1993) *The Human Genome Project* in association with MCET The Mass LearnPike, Cambridge, Massachusetts.
3. Brock T and Madigan M (1992) Gene manipulation and genetic engineering, *In: Biology of Microorganisms 6th Edition*. Englewood Cliffs, NJ: Prentice Hall, pp. 280-305.
4. Dodds JH (1985) *Plant Genetic Engineering*. New York: Cambridge University Press.
5. Goodman DC (1987) *From Farming to Biotechnology: A Theory of Agro-industrial Development*. Oxford: Blackwell.
6. Hurlbert RE (1999), *Genetic Engineering,* The Microbe Files, Washington State University.

7. Mulcahy H (1989) *A Simulated Laboratory Exercise in Genetic Engineering*. Abstracts of the Annual Meeting of American Society for Microbiologists.
8. Steinbrecher R (1999) *What is Genetic Engineering?* A Magazine of Green Social Thought, Vol. 18, Winter 1999, pp. 9-12.
9. Torrey JG (1987) The Development of Plant Biotechnology. American Scientist. 73:354-363.
10. Towle A (1988) Modern Biology. Austin, TX: Holt, Rinehart and Winston.Zhu Y. *On Genetic Engineering*. Vajra Bodhi Sea, April-May 1998.
11. Richard W (1988) *Genetic Transformation in Plants*, Milton Keynes, England: Open University Press.

3

Speaking the Language of Recombinant DNA

Recombinant DNA (rDNA) has various definitions, ranging from very simple to strangely complex. The following are three examples of how recombinant DNA is defined:

1. A DNA molecule containing DNA originating from two or more sources.
2. DNA that has been artificially created. It is DNA from two or more sources that is incorporated into a single recombinant molecule.
3. According to the National Institute of Health (NIH), US guidelines, recombinant DNA are molecules constructed outside of living cells by joining natural or synthetic DNA segments to DNA molecules that can replicate in a living cell, or molecules that result from their replication.

Recombinant DNA, also known as *in vitro* recombination, is a technique involved in creating and purifying desired genes. Molecular cloning (i.e. gene cloning) involves creating recombinant DNA and introducing it into a host cell to be replicated. One of the basic strategies of molecular cloning is to move desired genes from a large, complex genome to a small, simple one. The process of *in vitro* recombination makes it possible to cut different strands of DNA, *in vitro* (outside the cell), with a restriction enzyme and join the DNA molecules together via complementary base pairing.

Naturally Occurring DNA Recombination

The DNA sequence of a chromosome can change in large segments as well, by the processes of recombination and transposition. Recombination is the production of new DNA molecule(s) from two parental DNA molecules or different segments of the same DNA molecule.

At least four types of naturally occurring recombination have been identified in living organisms (Fig.1).

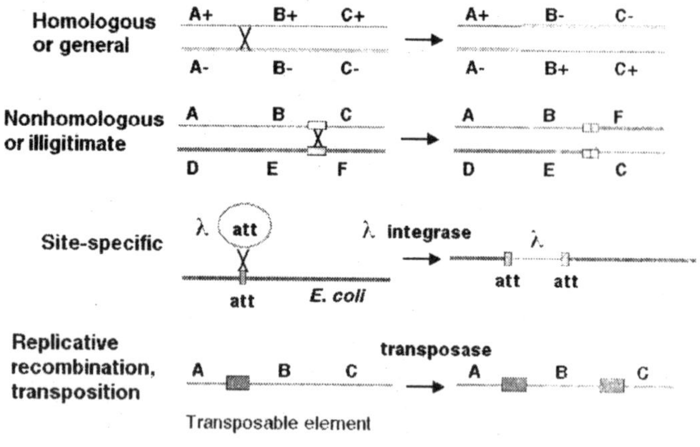

Figure 1. Types of naturally occurring DNA recombination. Each line represents a chromosome or segment of a chromosome; thus a single line represents both strands of duplex DNA.

General or homologous recombination occurs between DNA molecules of very similar sequence, such as homologous chromosomes in diploid organisms. General recombination can occur throughout the genome of diploid organisms, using one or a small number of common enzymatic pathways.

Illegitimate or nonhomologous recombination occurs in regions where no largescale sequence similarity is apparent, e.g. translocations between different chromosomes or deletions that remove several genes along a chromosome. However, when the DNA sequence at the breakpoints for these events is analyzed, short regions of sequence similarity are found in some cases. For instance, recombination between two similar genes that are several million base pairs (bp) apart can lead to deletion of the intervening genes in somatic cells.

Site-specific recombination occurs between particular short sequences (about 12 to 24 bp) present on otherwise dissimilar parental molecules. Site-specific recombination requires a special enzymatic machinery, basically one enzyme or enzyme system for each particular site. Good examples are the systems for integration of some bacteriophage, such as λ, into a bacterial chromosome and the rearrangement of immunoglobulin genes in vertebrate animals.

The fourth type is *replicative recombination*, which generates a new copy of a segment of DNA. Many transposable elements use a process of replicative recombination to generate a new copy of the transposable element at a new location.

Recombinant DNA technology uses two other types of recombination. The directed cutting and rejoining of different DNA molecules *in vitro* using restriction endonucleases and DNA ligases is well-known. Once made, these recombinant DNA molecules are then introduced into a host organism, often a bacterium. If the recombinant DNA is a plasmid, phage or other molecule capable of replicating in the host, it will stay extrachromosomal.

However, one can introduce the recombinant DNA molecule into a host in which it cannot replicate, such as a plant, an animal cell in culture, or a fertilized mouse egg. In order for the host to be stably transformed, the introduced DNA has to be taken up into a host chromosome. In bacteria and yeast, this can occur by homologous recombination at a reasonably high frequency. However, this does not occur in plant or animal cells. In contrast, at a low frequency, some of these introduced DNA molecules are incorporated into random locations in the chromosomes of the host cell. Thus random recombination into chromosomes can make stably transfected cells and transgenic plants and animals. The mechanism of this recombination during transformation or transfection is not well understood, although it is commonly used in the laboratory.

General recombination is an integral part of the complex process of meiosis in sexually reproducing organisms. It results in a crossing over between pairs of genes along a chromosome, which are revealed in appropriate matings. The chiasmata that link homologous chromosomes during meiosis are the likely sites of the crossovers that result in recombination.

General recombination also occurs in nonsexual organisms when two copies of a chromosome or chromosomal segment are present. We have encountered this as recombination during F-factor mediated conjugal transfer of parts of chromosomes in *E. coli*. Recombination between two phage during a mixed infection of bacteria is another example. Also, the retrieval system for post-replicative repair involves general recombination. The mechanism of recombination has been intensively studied in bacteria and fungi, and some of the enzymes involved have been well characterized. However, a full picture of the mechanism, or mechanisms, of recombination has yet to be achieved.

General recombination can appear to result in either an equal or an unequal exchange of genetic information.

Equal exchange is referred to as reciprocal recombination. In this example, two homologous chromosomes are distinguished by having wild type alleles on one chromosome (A+, B+ and C+) and mutant alleles on the other (A-, B-and C-). Homologous recombination between genes A and B exchanges the segment of one chromosome containing the wild type alleles of genes B and C (B+ and C+) for the segment containing the mutant alleles (B-and C-) on the homologous chromosome. This could be explained by breaking and rejoining of the two homologous chromosomes during meiosis.

The DNA that is removed from the top chromosome is joined with the bottom chromosome, and the DNA removed from the bottom chromosome is added to the top chromosome. This process resulting in new DNA molecules that carry genetic information derived from both parental DNA molecules is called reciprocal recombination. The number of alleles for each gene remains the same in the products of this recombination, only their arrangement has changed.

General recombination can also result in a one-way transfer of genetic information, resulting in an allele of a gene on one chromosome being changed to the allele on the homologous chromosome. This is called gene conversion. As illustrated in Fig.2, recombination between two homologous chromosomes A+B+C+ and A-B-C-can result in a new arrangement, A-B+C-, without a change in the parental A+B+C+.

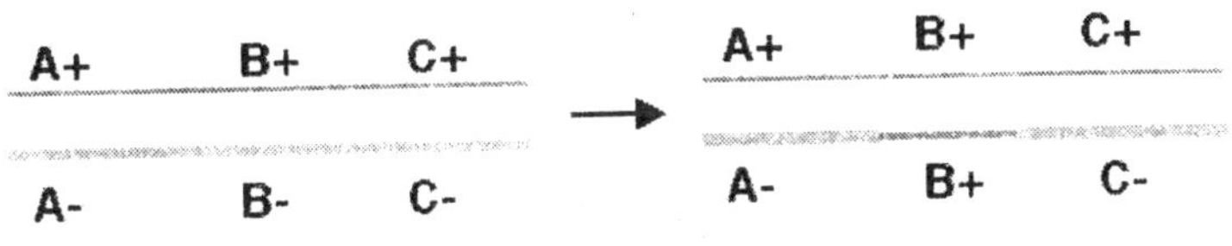

Figure 2. Gene conversion changing allele B-on the bottom chromosome to B+. Note that the arrangement of alleles on the top chromosome has not changed.

In this case, the allele of gene B on the bottom chromosome has changed from B-to B+ without a reciprocal change on the other chromosome. Thus, in contrast to reciprocal recombination, the number of types of alleles for gene B has changed in the products of this recombination; now there is only one (B+). This is an example of interchromosomal gene conversion, i.e. between homologous chromosomes. Similar copies of genes can be on the same chromosome, and these can undergo gene conversion as well. Cases of intrachromosomal gene conversion have been documented for the gamma-globin genes of humans. The occurrence of gene conversion during general recombination is one indication that the enzymatic mechanism is not a simple cutting and pasting.

Detecting Recombination

Mendel's Second Law described the random assortment of alleles of pairs of genes. However, certain pairs of genes show deviations from this random assortment, leading to the conclusion that those genes are linked on a chromosome. The linkage is not always complete, meaning that nonparental genotypes are seen in a proportion of the progeny. This is explained by crossing over between the gene pairs during meiosis in the parents.

Let's think about the general recombination shown in Fig. 1 in this context. The two chromosomes outlined in the figure are in a heterozygous parent, with the wild type alleles for genes A and B (A+ and B+) are on one chromosome and the mutant alleles (A-and B-) are on the homologous chromosome (We can ignore gene C for this discussion.) Homologous recombination during meiosis can generate the new chromosomes shown, now with A+ and B-on one chromosome and A-and B+ on the other. However, this crossover will not occur between genes A and B on all chromosomes undergoing meiosis in this parent. Although recombination is an essential part of meiosis, the sites of recombination on a particular chromosome varies from cell to cell. In fact, the probability that a crossover will occur between two genes is a measure of the genetic distance between them.

The recombinant chromosomes resulting from a crossover are revealed in a mating between the heterozygous parent (A+B+/A-B-) and a homozygous recessive individual (A-B-/A-B-). Most of the germ cells contributed by the heterozygous parent will have one of the parental chromosomes A+B+ or A-B-, but those germ cells resulting from the crossover between genes A and B will have the recombinant chromosomes (either A+B-or A-B+). The homozygous recessive parent will contribute only A-B-chromosomes.

Thus in the progeny, one sees mainly offspring whose phenotype is determined by one of the chromosomes in the heterozygous parent, either wild type A and B (genotype of A+B+/A-B-) or mutant A and B (genotype A-B-/A-B-). However, some of the progeny will show a wild type A and a mutant B phenotype, or vice versa. These carry the chromosomes resulting from the crossover (genotype of A+B-/A-B-or A-B+/A-B-). The frequency with which one sees progeny with nonparental phenotypes is related to their distance apart on the chromosome; this measure is referred to as a genetic distance or a recombination distance.

Meiotic Recombination

A diploid organism has two copies of each chromosome. If it has four chromosomes, there are two pairs, A and A' and B and B', not four different chromosomes A, B, C and D. One copy of each chromosome came from its father (e.g. A and B) and one copy of each came from its mother (e.g. A' and B'). Meiosis is the process of reductive division whereby a diploid organism generates haploid germ cells (in this case, with two chromosomes), and each germ cell has a single copy of each chromosome. In this example, meiosis does not generate germ cells with A and A' or B and B', rather it produces cells with A and B, or A and B', or A' and B, or A' and B'. The homologous chromosomes, each consisting of two sister chromatids, are paired during the first phase of meiosis, e.g., A with A' and B with B'.

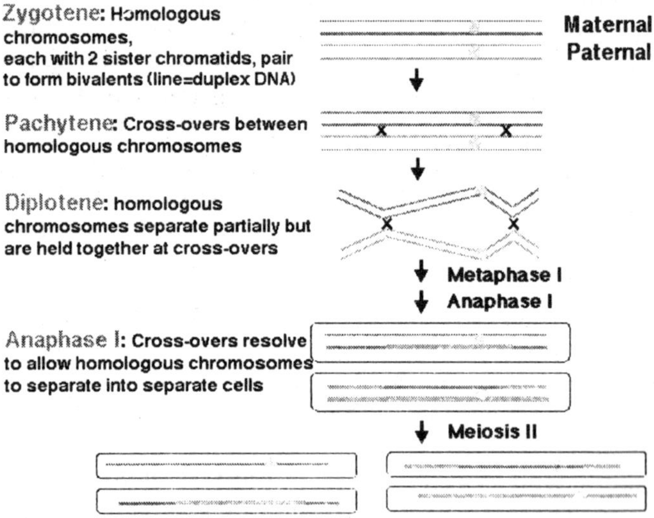

Figure 3. Homologous pairing and recombination during the first stage of meiosis (meiosis I). After DNA synthesis has been completed, two copies of each homologous chromosome are still connected at centromeres.

Then the homologous chromosomes are moved to separate cells at the end of the first phase, insuring that the two homologs do not stay together during reductive division in the second phase of meiosis. Thus each germ cell receives the haploid complement of the genetic material, i.e. one copy of each chromosome. The combination of two haploid sets of chromosomes during fertilization restores the diploid state, and the cycle can resume. Failure to distribute one copy of each chromosome to each germ cell has severe consequences. Absence of one copy of a chromosome in an otherwise diploid zygote is likely fatal. Having an extra copy of a chromosome (trisomy) also causes problems. In humans, trisomy for chromosomes 15 or 18 results in perinatal death and trisomy 21 leads to developmental defects known as Down's syndrome.

The ability of homologous chromosomes to be paired during the first phase of meiosis is fundamental to the success of this process, which maintains a correct haploid set of chromosomes in the germ cell. Recombination is an integral part of the pairing of homologous chromosomes. It occurs between non-sister chromatids during the pachytene stage of meiosis I (the first stage of meiosis) and possibly before, when the homologous chromosomes are aligned in zygotene (Fig. 3).

The crossovers of recombination are visible in the diplotene phase. During this phase, the homologous chromosomes partially separate, but they are still held together at joints called chiasmata; these are likely the actual crossovers between chromatids of homologous chromosomes. The chiasmata are progressively broken as meiosis I is completed, corresponding to resolution of the recombination intermediates. During anaphase and telophase of meiosis I, each homologous chromosome moves to a different cell, i.e. A and A' in different cells, B and B' in different cells in our example. Thus recombinations occur in every meiosis, resulting in at least one exchange between pairs of homologous chromosomes per meiosis.

Recent genetic evidence demonstrates that recombination is required for homologous pairing of chromosomes during meiosis. Genetic screens have revealed mutants of yeast and *Drosophila* that block pairing of homologous chromosomes. These are also defective in recombination. Likewise, mutants defective in some aspects of recombination are also defective in pairing. Indeed, the process of synapsis (or pairing) between homologous chromosomes in zygotene, crossing over between homologs in pachytene, and resolution of the crossovers in the latter phases of meiosis I (diakinesis, metaphase I and anaphase I) correspond to the synapsis, formation of a recombinant joint and resolution that mark the progression of recombination.

Advantages of Genetic Recombination

Not only is recombination needed for homologous pairing during meiosis, but recombination has at least two additional benefits for sexual species. It makes new combinations of alleles along chromosomes, and it restricts the effects of mutations largely to the region around a gene, not the whole chromosome.

Since each chromosome undergoes at least one recombination event during meiosis, new combinations of alleles are generated. The arrangement of alleles inherited from each parent are not preserved, but rather the new germ cells carry chromosomes with new combinations of alleles of the genes. This remixing of combinations of alleles is a rich source of diversity in a population.

Over time, recombination will separate alleles at one locus from alleles at a linked locus. A chromosome through generations is not fixed, but rather it is "fluid," having many different combinations of alleles. This allows nonfunctional (less functional) alleles to be cleared from a population. If recombination did not occur, then one deleterious mutant allele would cause an entire chromosome to be eliminated from the population. However, with recombination, the mutant allele can be separated from the other genes on that chromosome. Then negative selection can remove defective alleles of a gene from a population while affecting the frequency of alleles only of genes in tight linkage to the mutant gene.

Conversely, the rare beneficial alleles of genes can be tested in a population without being irreversibly linked to any potentially deleterious mutant alleles of nearby genes. This keeps the effective target size for mutation close to that of a gene, not the whole chromosome.

Recombinant DNA Techniques

Some of the molecular biology techniques utilized during recombinant DNA include:

1. *The study and/or modification of gene expression patterns*: Gene expression is the process by which a gene's coded information is converted into the structures present and operating in the cell. Expressed genes include those that are transcribed into mRNA (messenger RNA) and then translated into protein, and those that are transcribed into tRNA (transfer RNA) and rRNA (ribosomal RNA). Gene expression can be studied using microarray analysis, which is a method of visualizing the patterns of gene expression of thousands of genes using fluorescence or radioactive hybridization.
2. *Gene cloning*: Gene cloning utilizing recombinant DNA technology is the process of manipulating DNA to produce multiple copies of a single gene or segment of DNA.
3. *DNA sequencing*: DNA sequencing is a lab technique used to determine the sequence of nucleotide bases in a molecule of DNA.
4. *Creation of transgenic plants and animals*: A transgenic plant or animal is one who has been genetically engineered, and usually contains genetic material from at least one unrelated organism, such as from a virus, other plant, or other animal.

Process of Making Recombinant DNA

The following is a summary of the process of making recombinant DNA:

1. Treat the DNA taken from both sources with the same restriction endonuclease.

2. The restriction enzyme cuts both molecules at the same site.

3. The ends of the cut have an overhanging piece of single-stranded DNA called "sticky ends."

4. These sticky ends are able to base pair with any DNA molecule that contains the complementary sticky end.

5. Complementary sticky ends can pair with each other when mixed.

6. DNA ligase is used to covalently link the two strands into a molecule of recombinant DNA.

7. In order to be useful, the recombinant DNA needs to be replicated many times (i.e. cloned). Cloning can be done *in vitro*, via the Polymerase Chain Reaction (PCR), or *in vivo* (inside the cell) using unicellular prokaryotes (e.g. *E. coli*), unicellular eukaryotes (e.g. yeast), or mammalian tissue culture cells.

Examples of therapeutic products made by recombinant DNA

Some examples of the therapeutic products made by recombinant DNA techniques include:

1. Blood proteins: Erythropoietin; Factors VII, VIII, IX; Tissue plasminogen activator; Urokinase

2. Human Hormones: Epidermal growth factor; Follicle stimulating hormone; Insulin; Nerve growth factor; Relaxin; Somatotropin

3. Immune modulators: α-Interferon; β-interferon; Colony stimulating factor; Lysozyme; Tumor necrosis factor

4. Vaccines: Cytomegalovirus; Hepatitis B; Measles; Rabies

Cloning Vectors

A cloning vector is a DNA molecule that carries foreign DNA into a host cell (usually bacterial or yeast), where it replicates, producing many copies of itself along with the foreign DNA.

There are three features required for all cloning vectors:

1. Sequences that will permit the propagation of itself in bacteria or in yeast.
2. A cloning site to insert the foreign DNA; the most versatile vectors contain a site that can be cut by many restriction enzymes.
3. A method of selecting for bacteria or yeast containing a vector with foreign DNA; this is usually accomplished by selectable markers for drug resistance

General Steps of Cloning with Vectors

1. Prepare the vector and the DNA to be cloned by digestion with restriction enzymes to generate complementary ends.
2. Ligate (join) the foreign DNA into the vector with the enzyme DNA ligase.
3. Introduce the DNA into bacterial cells or yeast cells by transformation.
4. Select cells containing foreign DNA by screening for selectable markers (commonly drug resistance).

Types of Cloning Vectors

— Plasmid

— Phage

— Cosmid

— Bacterial Artificial Chromosomes (BAC)

— Yeast Artificial Chromosomes (YAC)

Plasmids

Plasmids are small, circular, extrachromosomal DNA molecules found in bacteria, which can replicate on their own, outside of a host cell (Fig. 4). They have a cloning limit of 100 to 10,000 base pairs or 0.1-10 kilobases (kb). A plasmid vector is made from natural plasmids by removing unnecessary segments and adding essential sequences. Plasmids make excellent cloning vectors for various laboratory techniques, including recombinant DNA.

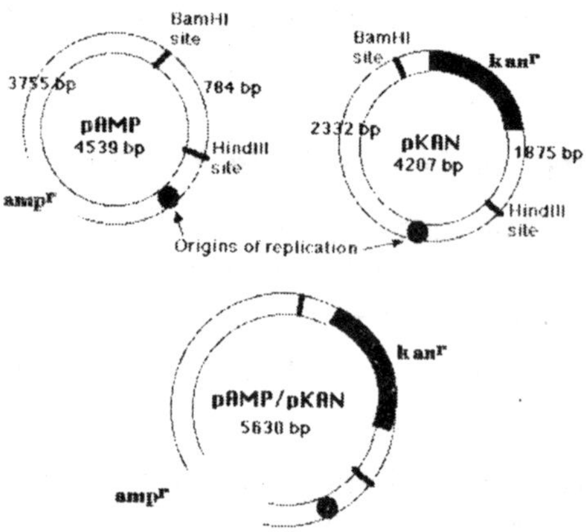

Figure 4. Plasmids

Transformation is the modification of the genotype of a cell (usually prokaryotic) by introducing DNA from another source. During transformation, genetic information is transferred via the uptake of free DNA. Often these sources of DNA come from plasmids that are deliberately introduced into a cell, transforming the cell with its genes. Transformation occurs naturally, and the resulting uptake of foreign DNA by the cell is not typically considered recombinant DNA. If the plasmid being introduced to the cell has not been genetically altered, *in vitro*, then the plasmid is not considered to contain recombinant DNA.

Plasmids that have been genetically altered, such as by the insertion of an antibiotic resistance gene, do contain recombinant DNA because the original genetic composition of the plasmid has been artificially altered. These plasmids can be used to incorporate bacterial cells with the antibiotic resistance gene via transformation.

Phages

Phages are derivatives of bacteriophage lambda (λ phage), a virus which infects *E. coli*. They are linear DNA molecules. whose region can be replaced with foreign DNA without disrupting its life cycle. The major advantage of the λ phage vector is its high transformation efficiency, which is about 1000 times greater than that of the plasmid vector. They also have a larger cloning limit than plasmids, consisting of 8-25 kb (Fig. 5).

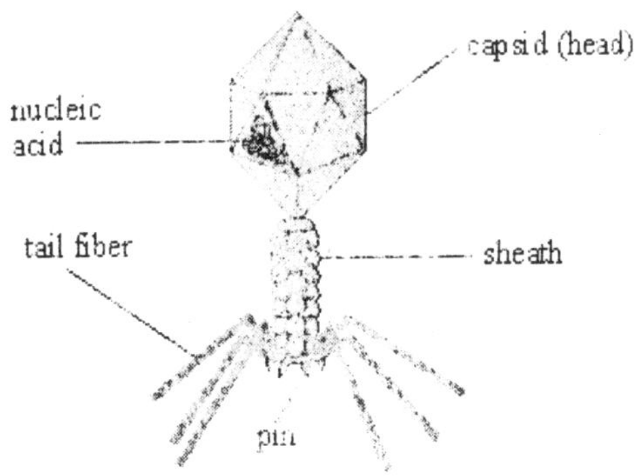

Figure 5. Bacteriophage

Cosmids

Cosmids are extrachromosomal circular DNA molecules that combine features of plasmids and phages (Fig. 6). They also have a high transformation efficiency, and their cloning limit of 35-50 kb is larger than that of plasmids or phages.

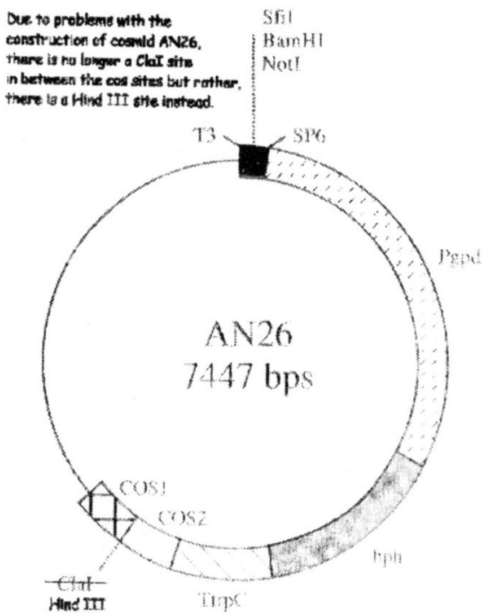

Figure 6. Cosmids

Bacterial Artificial Chromosomes (BAC)

BACs are based on bacterial mini-F plasmids, which are small pieces of episomal bacterial DNA that give the bacteria the ability to initiate conjugation with adjacent bacteria. They have a cloning limit of 75-300 kb.

Yeast Artificial Chromosomes (YAC)

YACs are artificial chromosomes that replicate in yeast cells. They consist of:

- — Telomeres, which are ends of chromosomes involved in the replication and stability of linear DNA.
- — Origin of replication sequences necessary for the replication in yeast cells.
- — A yeast centromere, which is a specialized chromosomal region where spindle fibers attach during mitosis.
- — A selectable marker for identification in yeast cells.
- — Ampicillin resistance gene for selective amplification.
- — Recognition sites for restriction enzymes.

The procedure for making YAC vectors is as follows:

1. The target DNA is partially digested by a restriction endonuclease, and the YAC vector is cleaved by restriction enzymes.
2. The cleaved vector segments are ligated with a digested DNA fragment to form an artificial chromosome.
3. Yeast cells are transformed to make a large number of copies.
4. They are the largest of the cloning vectors, with a cloning limit of 100-1000 kb, however they have very low efficiency.

Yeast

Yeasts, eukaryotic unicellular fungi, contribute a great deal to the study of molecular genetics. They are popular organisms to clone and express DNA in because they are eukaryotes, and can therefore splice out introns, the non-coding sequences in the middle of many eukaryotic genes.

For the past two decades *Saccharomyces cerevisiae*, a species of yeast, has been an important model system for biological research because its entire genome has been base sequenced, and is used as a reference to human and other higher eukaryotic genes. This is because the basic cellular mechanics of replication, recombination, cell division and metabolism are generally conserved between yeast and larger eukaryotes, including mammals. Also, yeast is easily genetically manipulated, which allows for convenient analysis and the functional dissecting of gene products from other eukaryotes.

Gene Therapy

Gene therapy can be defined as the transfer of genetic material into the cells of an individual resulting in a therapeutic benefit to the individual. It involves the intentional modification of genetic material with the aim of preventing, diagnosing, or curing a disease. These modifications include the correction of a genetic defect resulting from the absence or alteration of a protein, or the addition of genetic information to modify cellular characteristics.

Gene therapy allows the modification of specific genes without having to alter the disease phenotype using agents that either interact with the gene products (proteins), or are gene products themselves. The genetic modifications can be done *in vitro* or directly *in vivo* by using vectors that are capable of genetic transfer. Products used in gene therapy include viral vectors, genetically modified cells, and free or complex nucleic acids.

Currently the most efficient method for gene transfer is viral vectors. They cause infection of the target cell and transfer the therapeutic gene using their natural biological mechanisms. Viral vectors used in gene therapy must be unable to replicate and have no lytic (ability to rupture a cell) activity. The result of gene therapy is the permanent treatment of disease, hopefully with few or no side effects in the process.

Viral Vectors

A viral vector is a virus that carries a modified or foreign gene. They are commonly used in gene therapy where the viral vector delivers the desired gene to a target cell.

Some of the viruses used as vectors in gene therapy include:

— Retroviruses

— Adenoviruses

— Parvoviruses

— Herpesviruses

— Poxviruses

Retroviruses

Retroviruses are viruses belonging to the family Retroviridae. They are composed of a single RNA strand and use the enzyme reverse transcriptase to copy their genome into the DNA of the host cell's chromosomes (Fig. 7). They are relatively genetically simple, and have the ability to infect a wide variety of cell types with high efficiency.

Retroviruses are the cause of many infections and cancers in vertebrates, including human T-cell leukemia and HIV. They also cause a variety of hematopoetic and neurological conditions, including paralysis, wasting, ataxia, arthritis, dementia, and neuropathy. Although vertebrate infection is mainly focused on, retroviruses have also been identified in virtually all organisms including invertebrates. The genome of all retroviruses consists of two molecules of RNA, which are single stranded, (+) sense, and have 5' cap and 3' poly-(A) (which is equivalent to mRNA). They vary in size from approximately 8 to 11 kb. Retrovirus genomes have 4 unique features:

1. They are the only viruses that are truly diploid (i.e. contain a full set of genetic material which consists of paired chromosomes).
2. They are the only RNA viruses whose genome is produced by cellular transcriptional machinery, i.e. they do not require participation by a virus-encoded polymerase.
3. They are the only viruses whose genome requires tRNA for replication.
4. They are the only (+) sense RNA viruses whose genome does not serve directly as mRNA immediately after infection.

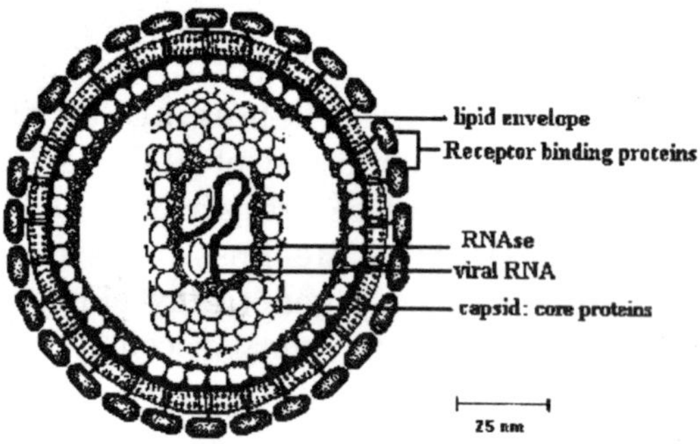

Figure 7. Retrovirus

Avian and murine retroviruses are commonly used as cloning vectors in gene therapy. Lentiviruses are a sub-group of retroviruses that are able to infect non-dividing cells or terminally differentiated cells, such as neurons. The main lentivirus vectors are derived from immunodeficiency viruses, including HIV.

Adenoviruses

Adenovirus can be defined as a group of DNA containing viruses, which most commonly cause respiratory disease (ranging from one form of the common cold to pneumonia, croup, and bronchitis), and can also cause gastrointestinal illness, eye infections, cystitis, and rash in humans. Adenoviruses can also be genetically modified and used in gene therapy to treat cystic fibrosis, cancer, and potentially other diseases.

Some of the characteristic features of adenoviruses include:

— They are widespread in nature, infecting birds, and many mammals including humans. There are two genera: Aviadenovirus (avian) and Mastadenovirus (mammalian)

— They can undergo latent infection in lymphoid tissues, and become reactivated at a later time.

— Several types have oncogenic (cancer-causing) potential.

Adenoviral vectors can be used to express a wide variety of viral and cellular genes in mammalian cells. They can also be used to transfect cells for protein expression and *in vivo* characterization studies. Transfection is the uptake, incorporation, and expression of recombinant DNA by eukaryotic cells.

Adenoviral vectors are great to use for transfection because the adenovirus efficiently infects many different cell lines. The virus enters the cell but it does not replicate because the essential E1 (viral genes which are expressed early in the viral life cycle) are absent. This is termed abortive infection, and it can be used as a transfection system for introducing a functional gene into cells. Adenoviral vectors are used in gene therapy when a high level of expression of the transferred gene is required for brief periods of time. The advantages of using adenovirus for the introduction of genes into cells include:

1. The process is quick, simple, and does not require any special equipment.

2. It is well tolerated, contrary to other methods. Post-infection viability is near 100% which means adenovirus can transfer genetic material into cells without toxic effects.
3. Adenovirus can infect most cell types.
4. Gene expression can be analyzed as early as a few hours after infection.
5. More than one protein can be expressed simultaneously.

Adenoviruses are medium-sized, nonenveloped, regular icosahedrons of 65-100 nanometers in diameter. The genome of Adenoviruses is a linear double-stranded DNA molecule containing 30,000-42,000 nucleotides.

Adenoviruses are good vectors for use in human gene therapy because it can infect cells *in vivo*, rather than manipulating the cells *in vitro*, and returning the cells to the body. In order to become a vector, the genome of the adenovirus must first be altered by removing all the viral DNA, except for the minimum necessary for the virus to live and infect the cells.

Genetically engineered viral vectors like these are harmless and typically can not live outside of the laboratory. A copy of the Cystic Fibrosis gene is then inserted into the viral genome via *in vitro* recombination. The viral genome is now considered recombinant. Then purified proteins and enzymes are used to build a new virus, or "package" the DNA. To do this, the recombinant DNA is mixed with all of the protein components of the virus along with some viral enzymes that assemble the virus, and in the test tube a whole, intact virus is created with the new DNA inside. After infecting some cells in the laboratory, the new virus makes millions of copies of itself (carrying the Cystic Fibrosis gene), which can be purified and used for gene therapy.

Parvoviruses

Parvoviruses are small DNA viruses that cause several diseases in mammals, such as canine parvovirus in dogs. Parvovirus B-19, which causes Fifth disease (erythema infectiosum) in humans, is the only form that is pathogenic to humans. In fact, many parvoviruses exhibit oncosuppressive properties (suppression of cancer-causing genes). Parvovirus-based vectors can be used to target the expression of therapeutic genes in tumors.

Herpesviruses

Herpesviruses include infectious human viruses including herpes simplex virus type-1 (HSV-1), which is common in the general population, but in rare cases can cause encephalitis. It is one of the most commonly used vector systems because it has a broad host cell range, the ability to transduce neurons, and a capacity to receive large inserts (Fig. 8).

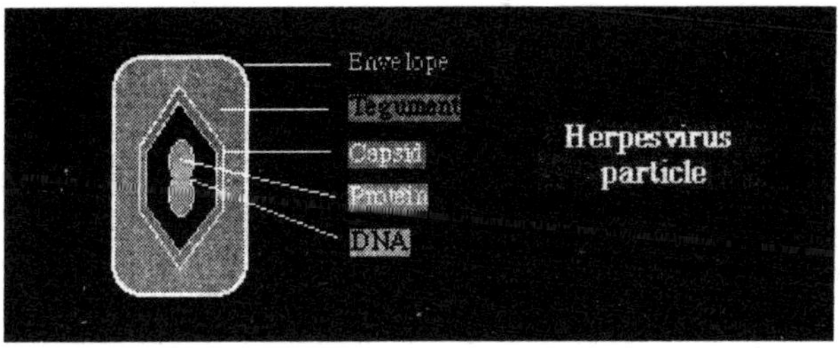

Figure 8. Herpesvirus

Poxviruses

Poxviruses are the largest and most complex viruses known. There are at least nine different poxviruses that are pathogenic to humans, the most common being vaccinia and variola virus (smallpox), which has since been erradicated. Poxviral vectors are successful in immunogenetic therapy protocols, due to their strong immnuogenicity (ability to induce a high immune response). They cause the activation of immune responses against tumor antigens transported in dendritic cells.

Enzymatic Manipulation of DNA and RNA

Restriction Endonucleases

Restriction endonucleases are bacterial enzymes that cut double-stranded DNA in a sequence-specific manner. Biologically, the function of these enzymes is to "restrict" the entry of foreign DNA by cleaving it at recognition sites that do not exist in the host bacterium. Each enzyme has a specific sequence that it recognizes, binds, and cuts. These recognition sites usually consist of a code of 4-8 bp, and the restriction endonucleases will only cut sequences that perfectly match this code. For example, the recognition sequence for the enzyme EcoRI is GAATTC. All sites with this sequence will be cleaved by the enzyme (Fig. 5), but even a single base mismatch, such as CAATTC, will prevent DNA digestion. The chance of finding a specific restriction site in a given piece of DNA is largely dependent on the number of bases in the recognition site. Based on the probability that the four nucleotides (GATC) will create a characteristic restriction site, a 4-base cutter should cut human genomic DNA every 2.5×10^2 bp, whereas an 8-base cutter would cut on average of every 1.0×10^6 bp.

Restriction enzymes can cut DNA in three basic ways (Fig. 9): they can cut the recognition site asymmetrically to leave ends with 5' or 3' overhangs, or they can cut symmetrically to leave blunt ends. Restriction enzymes are named according to the bacteria from which they were isolated (e.g., EcoRI from Escherichia coli). Hundreds of different restriction endonucleases are commercially available today.

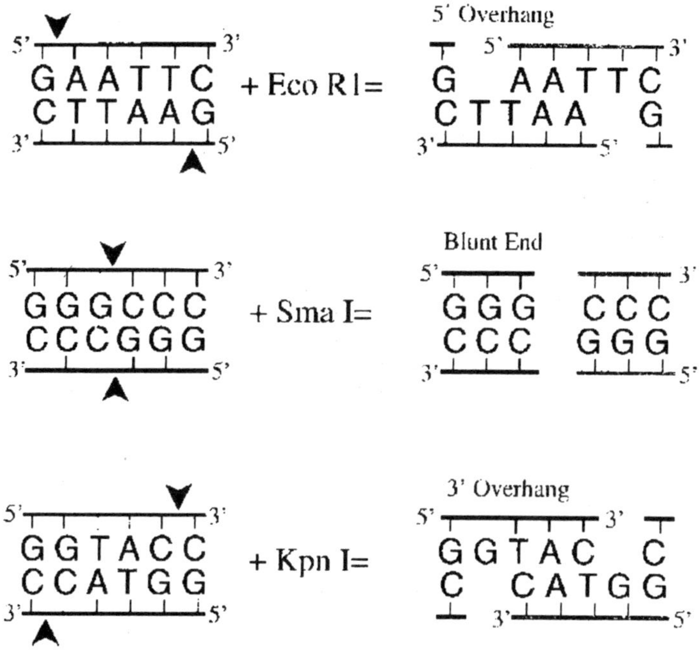

Figure 9. Restriction enzyme cleavage. Three enzymes are shown which digest DNA in three separate ways. EcoRI, 5' overhang; SmaI, blunt end; KpnI, 3' overhang. Arrowheads indicate the site of cleavage within the restriction site for each enzyme.

In practical terms, the restriction enzyme is incubated with DNA in a compatible buffer solution and cuts the DNA into pieces of a defined size. Because the cleavage is sequence-specific, different DNAs will produce a characteristic digestion pattern. These unique DNA fingerprints can be recognized after separation of the different size fragments by gel electrophoresis. The production of such restriction maps represents one of the major uses of these enzymes. A second major use of restriction endonucleases is to "cut and paste" DNA in order to generate novel DNA sequences that are used in a variety of methods.

DNA ligases

DNA ligases are enzymes that attach two pieces of DNA covalently to one another. T4 DNA ligase is the most commonly used enzyme. The reaction occurs between the 3' hydroxy group of one DNA strand and the phosphate group of the partner strand. ATP provides the energy for the reaction.

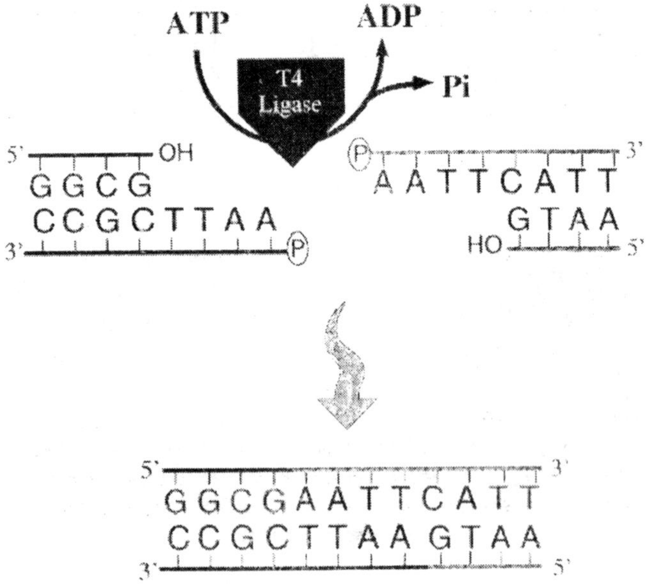

Figure 10. Ligase reaction. T4 DNA ligase will catalyze the joining of complementary ends of DNA by joining the 5' phosphate group to the 3' OH group of two strands of DNA. This reaction requires ATP as well as compatible ends of DNA.

DNA ligation can occur efficiently only between compatible ends of DNA fragments, and this specificity gives the investigator some control over how different DNA fragments ligate together in a reaction mix. The most common use of ligation is to incorporate a piece of DNA into a cloning vector (Fig. 10).

Reverse transcriptase (RT)

Reverse transcriptase (RT) enzymes are a subtype of DNA polymerase that utilize RNA as a template to produce DNA. They also require a primer and synthesize in the 5' → 3' direction. Reverse transcriptases are isolated from retroviruses, which have the ability to reverse copy their RNA genome for their life cycle in mammalian cells. The major use of RT is to copy mRNA into DNA, which is then called "complementary DNA" or cDNA. The reaction of RT on an RNA template produces a hybrid RNA-DNA molecule. The DNA in this hybrid is called "first-strand cDNA" and can be used to generate another complementary copy of DNA to create double-stranded DNA (dsDNA). In this manner, cDNA is used to generate libraries representing the expressed genes of a tissue and as a substrate for polymerase chain reaction (PCR) in the sequence analysis of RNA species.

Radioactive DNA

Radioactive DNA is an important reagent in molecular biology, as it serves as a valuable hybridization probe. There are two basic ways to incorporate radioactivity into DNA: internal labeling and end-labeling. For internal labeling, DNA is denatured into single strands by boiling, and DNA polymerase is used to copy the strand in the presence of a radioactively labeled nucleotide (most commonly 32P-dCTP). The newly synthesized strand is therefore highly radioactive. Polynucleotide kinase is used for end-labeling. This enzyme transfers the radioactivity from an ATP molecule to the 5' end of single-stranded DNA. Only the most 5' base is labeled, and no radioactivity is incorporated internally in the DNA molecule. Several nonradioactive methods are also available for "labeling" DNA. For example, biotinylation allows detection with avidin-based systems, and digoxygenin can be detected with enzyme-linked assays.

Isloation of DNA and RNA

In order to be accessible for manipulation, diagnostic studies, and therapeutic uses, DNA and RNA must be isolated in suitable form and quantity. The best method of DNA or RNA isolation, and propagation depends on the intended use. Although most techniques are labor-intensive and require large amounts of starting material, the use of isolation kits from a variety of biotechnology companies have made the isolation of DNA and RNA a relatively simple matter. Both classical isolation and kit isolation methods utilize the basic principles outlined below.

Isloation of Total Genomic DNA

Genomic DNA is double-stranded and can be isolated from any kind of cell, including bacteria, yeast, plants, and animal cells. For most uses, it is desirable that the DNA remain as undamaged and the highest molecular weight as possible. The process of purifying DNA tends to introduce strand breaks and to reduce its molecular weight, but proper precautions can minimize shearing. There are many variations in the details of DNA isolation protocols, but all involve lysis of the cell and nucleus (if the cell has a nucleus) as a first step. A suspension of the DNA-containing cells is incubated with detergents and/or lytic enzymes for this purpose.

The next step involves dissociation of the DNA from any attached proteins (such as histones). Most protocols utilize a strong protein degrading enzyme (proteinase K) for this purpose. This step also inactivates any DNA degrading cellular enzymes (nucleases). The partially degraded proteins are removed from the solution by phenol/chloroform extraction, salt precipitation, or the solution being passed over a column, which retains the DNA. DNA is precipitated from the remaining solution by ethanol precipitation. High-molecular-weight DNA will form threads when precipitated in this way and can then be "spooled" from the solution with a glass rod. The DNA is then dried and redissolved in an aqueous buffer for further use and is stable in this form for several years.

Isloation of Total Cellular DNA

Several different forms of RNA exist in cells, including messenger RNAs (mRNA), ribosomal RNAs (rRNA), transfer RNAs (tRNA), and others. For many types of studies (Northern blots, RT-PCR), the isolation of total cellular RNA species is acceptable. In contrast to DNA, RNA is single-stranded and relatively unstable. During RNA isolation and storage, constant caution should be taken to avoid degradation by RNase enzymes, which are ubiquitous.

For RNA isolation, cells, or tissues are homogenized in a solution that contains RNase-inhibiting compounds, such as the chaotropic chemical guanidinium isothiocyanate. Many different methods are available for the next step, which involves separating the RNA from genomic DNA. As in DNA isolation, the purified RNA is precipitated with ethanol during the final step of all procedures. Aqueous RNA solutions should be stored frozen.

Isolation of mRNA

For certain applications, total cellular RNA is inadequate. Ribosomal RNA represents the bulk of total cellular RNA and only 1-3% of all RNA is mRNA. In order to make libraries of expressed sequences or to detect rare mRNA species on Northern blots, it is important to isolate mRNA as the starting material. Most mRNAs are poly-adenylated (poly[A])RNA, and this feature can be exploited for their capture. Total cellular RNA is passed over a column or resin of polydeoxythymine (oligo-dT column, poly-dT) bound to a solid phase. Poly(A) RNA binds to the poly-dT and is immobilized on the solid phase. The remaining unbound RNA is then washed off, and the poly(A) RNA is eluted in a subsequent step.

DNA Cloning

In the DNA isolation method described above, the genome of cells is the source of DNA; therefore, all sequences will be represented in the same abundance as they exist in the cell. Depending on the genome size of the organism, individual genes of interest represent only a minute fraction of the total DNA in a cell. For example, the average single gene in humans (~100,000-bp size) represents only 0.003% of the total genome. Thus, individual genes are much too dilute in such samples to be useful for detailed studies (such as DNA sequencing).

This problem is solved by the cloning of DNA, which allows the preparation of large and pure quantities of a single DNA species. All methods of DNA cloning involve the ligation of a sequence of interest (such as a human gene) into a cloning vector. Cloning vectors are DNA sequences that exist inside a cell (in most cases the bacterium *E. coli*) but are not part of the genome of that cell. The cloning vehicle (and the cloned DNA within it) replicates inside the host cell, producing large quantities of a specific DNA sequence that can be isolated for further study. Cloning into *E. coli* plasmids remains the most important method for preparing large quantities of pure DNA (Fig. 11). Plasmids are small circles of double-stranded DNA (2000-5000 bp) that replicate inside bacterial cells.

Plasmids have their own origin of replication and can propagate themselves with the help of the endogenous cellular machinery. Cloning plasmids contain a selectable marker, usually a gene that confers resistance to an antibiotic. *E. coli* cells containing this antibiotic-selectable plasmid are then able to grow in media containing the respective antimicrobial agent.

Many plasmids contain the b-lactamase gene, allowing growth in ampicillin. Low-and high-copy-number plasmids are distinguished. Pure plasmid DNA can be isolated from *E. coli* genomic DNA, because plasmid DNA is much smaller in size than *E. coli* genomic DNA. The plasmid-containing *E. coli* cells are grown in a large volume of liquid culture media containing the antibiotic of choice. The cells are then pelleted by centrifugation and lysed, either with alkali, or using detergents.

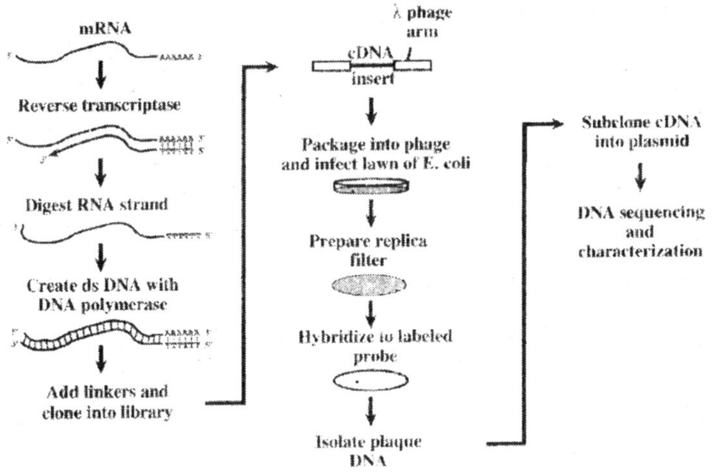

Figure 11. Subcloning in bacteria. pUC 19 cloning vector is used propagate cDNA. The plasmid carries a bacterial origin of replication (ORI). Also, a resistance gene for selection in ampicillin media (Amp), as well as a gene that can be used to detect for insertion cDNA (lacZ). After opening the plasmid and introduction of the cDNA through restriction enzyme digestion and ligation, the construct is transformed into bacteria (E. coli) through either heat shock or electroporation. After transformation, bacteria are plated on resistant media, allowing the bacteria to grow only if they have taken up the plasmid containing the resistance gene. Growing the colonies with X-gal, a substrate for B galactosidase, clones that have been inserted into the lacZ gene and disrupted its reading frame, will no longer be able to metabolize X-gal and will appear white on the media plate. However, those colonies that may have religated and still contain a functional lacZ gene will metabolize X-gal and have a blue appearance.

All plasmid preparation methods use size selection to remove the high-molecular-weight genomic DNA and RNA leaving only the small circular plasmid DNA. Milligram quantities of pure plasmid DNA can be prepared from a 1-L culture of *E. coli* cells containing a high-copy-number plasmid. Modern *E. coli* plasmids have been engineered to contain a region with multiple useful restriction endonuclease sites (polylinker sites). In order to clone a gene of interest, the following procedure is followed :

DNA from the parental plasmid is cut with a restriction endonuclease, thus opening the ring and converting it to a linear molecule. A linear double-stranded DNA sequence of interest is generated with matching ends and is then ligated into the gap to regenerate a circular plasmid. The ligated plasmid is introduced by a process called transformation into a strain of *E. coli* that is sensitive to the antibiotic resistance marker carried by the plasmid. Transformation of bacteria can be accomplished by either heating the DNA-bacteria mix at 42°C for 1 min or by a quick pulse of electricity (electroporation).

Both methods move the plasmid DNA through the bacterial wall and into the bacterial cell. After transformation, the bacteria are grown on antibiotic-containing plates. Only cells that have taken up a circular plasmid are able to survive and form colonies in the presence of the antibiotic. Individual bacterial colonies are also called "clones," because every cell in the colony is genetically identical. Individual clones can be expanded in liquid media for the preparation of large amounts of plasmid DNA. Since bacterial cells are immortal and can continue to divide without limit, plasmid-containing clones represent a permanent source of DNA. The cells can be stored frozen indefinitely, and clones can be mailed and shared with other laboratories.

Modern high-copy-number plasmids are about 3 kb in size and can accommodate inserts of up to 15 kb. Most DNA cloned in plasmids, however, is less than 10 kb in size and larger sequences are frequently unstable. Thus, only relatively small pieces of DNA can be cloned in plasmid vectors. For this reason, multiple cloning vectors capable of harboring larger DNA sequences have been developed.

Phage Cloning Vectors

Bacteriophage λ is a virus that infects *E. coli* cells and has been used extensively for cloning. Its structure is very different from plasmids. Virus particles consist of a protein envelope and core and a linear double-stranded DNA genome. For cloning purposes, much of the wild-type genome has been removed to allow replacement by exogenous DNA. In contrast to plasmids, inserts of up to 23 kb in size are stable in phage.

Cloning in phage is performed in the following way:

A large internal portion of the wild-type phage genome is removed by restriction digestion leaving a long and short arm. Neither of these two arms alone contains enough sequence to produce an infectious phage nor are the arms large enough to be packaged. The DNA sequence of interest is then ligated between the two arms, restoring a single large DNA genome. A "packaging extract" containing viral structural proteins isolated from wildtype phage is then used to package the recombinant phage and to infect *E. coli* cells.

After infection of a bacterial cell, λ phage replicates to high copy numbers within the cell and then lyses it. Phage particles are released and can infect other *E. coli* cells. When propagated on agar plates, phage will then lyse bacterial cells and leave a hole or "plaque" in the lawn of bacteria. Each plaque represents a clone of DNA, because all the phage in a plaque are genetically identical. When a liquid culture of *E. coli* is infected with phage, most of the cells will lyse and a high density of phage particles is attained. Phage particles can then be purified and their DNA isolated in large quantities.

Recently, one other *E. coli* bacteriophage has become an important cloning vehicle. Phage P1 can stably harbor DNA inserts of up to 100 kb and is therefore used to clone long stretches of genomic DNA.

Cosmids

Cosmids harbor inserts of ~40 kb and are very useful cloning vectors. Their properties are similar to plasmids in many aspects. Cosmids can propagate in *E. coli* and do not lyse their host cell. Antibiotic resistance genes (usually ampicillin) are used to select for cosmid-containing cells, which form colonies on plates. Inserts are ligated into convenient restriction sites.

Bacterial Artificial Chromosomes

Bacterial artificial chromosomes (BACs) are a special *E. coli* plasmid that can stably replicate very large inserts of genomic DNA. BAC inserts can be as large as 180 kb, and complete libraries of the human genome are now available in BACs. In contrast to YACs, BAC inserts are usually pure (nonchimeric), and the BAC DNA is very easy to isolate.

Yeast Artificial Chromosomes (YACs)

Yeast artificial chromosomes (YACs) have become very important in the human genome project, because they can harbor by far the largest inserts (up to 2000 kb). Because of this, a single YAC can cover a relatively large genomic region. In contrast to all other cloning vehicles described here, YACs are not grown in bacteria, but in the yeast *Saccharomyces cerevisiae*. YACs consist of a artificial centromere, the insert, and an artificial telomere, thus literally representing an artificial chromosome. Both the artificial centromere and the telomere contain different selectable marker genes.

In order to generate a YAC, large pieces of DNA are ligated to the centromere on one end and the telomere on the other. This construct is then introduced into yeast, which is incubated on selective media plates. Only those yeast cells that contain both the telomere and centromere (and therefore both selectable marker genes) will survive and form colonies. Individual colonies (clones) can then be picked, grown in liquid culture, and YAC DNA can be isolated from the yeast cells.

YACs have the advantage of containing very large inserts, but they have the disadvantage of frequently being chimeric. Chimeric YACs contain inserts from more than one region of the genome, which can limit their experimental utility. For example, a chimeric YAC may contain an insert from chromosome 3p fused to a portion of chromosome 15.

DNA Libraries

Cloning vectors such as those described above can be used to replicate a DNA sequence of interest and purify large quantities. But how are genes of interest found in the first place? The process by which gene(s) are isolated from the complex mix of DNA sequences found in cells of higher organisms will be described briefly.

Most gene-isolation strategies have traditionally involved the generation of a library of DNA or RNA sequences. The beginning material for the library is a complex mix of DNA known to contain the sequence of interest. For example, the starting material for the isolation of a human gene would be total genomic DNA isolated from human cells. This DNA is then broken down into fragments of the appropriate size by mechanical shearing or by restriction endonuclease digestion.

The appropriate size depends on the cloning vector to be used. For λ phage libraries, the average fragment size should be 8-25 kb. This mixture of fragments is ligated into a cloning vector and the ligation reaction is used to infect or transform *E. coli* bacteria. The biology of cloning vectors assures that each individual bacterium will contain only one cloning vector with an insert. Thus, this procedure divides the initial complex mix of DNA into millions of individual clones, each of which contains only a single DNA sequence, a library. Libraries can be produced from any source of double-stranded DNA.

In order to identify the protein-coding regions of genes, however, it is necessary to generate libraries from expressed sequences. This type of library is called cDNA library and the starting material is messenger RNA (Fig. 8). To isolate the gene for a liver metabolic enzyme, for example, mRNA would be purified from homogenized liver. This material contains all mRNAs that are present in the liver. Some mRNAs will be very abundant, because they encode proteins present in high amounts in liver. Others transcripts will be rare. Reverse transcription is then used to copy the mix of mRNAs into first-strand cDNA. The cDNAs are then converted to double-stranded DNA and ligated into a cloning vector to produce the cDNA library. It is important to note that cDNA libraries are tissue-specific, because each tissue expresses a different set of genes. Similarly, cDNA libraries are specific for a developmental stage of the organism.

Once a library of clones has been generated from the nucleic acid source of interest, the challenge is to find the specific individual clone with the gene among the millions of clones in the library. This is achieved by library screening. The bacteria containing the library are spread on agar plates at a defined density, which is low enough to permit the isolation of individual clones.

A phage library produced from human genomic DNA, for example, would be plated a density of 50,000 plaques per plate. At this density, 20 plates represent about 1,000,000 individual clones. Part of each colony/plaque on a plate is then transferred, as an exact copy, onto a nitrocellulose or nylon filter in a procedure termed "filter lift." These filters are then processed to covalently bind either the DNA or the expressed proteins of the transferred plaques/colonies.

The filters are incubated with a radioactively labeled DNA probe or specific antibodies, which will bind to the position on the filter occupied by a individual clone containing the desired DNA or expressing the sought-after protein. The corresponding clone is then picked from the original plate based on the position marked on the filter. DNA in the clone can then by amplified and isolated as described above, usually by subcloning it into plasmid vectors.

Expression of Recombinant Proteins

The ability to express large quantities of functional recombinant proteins is one of the many advantages conferred by the cloning of cDNAs. There are many practical uses for recombinant proteins (Table 1). They provide unlimited sources for reagents such as restriction enzymes and represent an invaluable source for hormones and growth factors, such as human insulin and erythropoietin, that are difficult to isolate in abundance from natural sources. The availability of abundant amounts of recombinant proteins has greatly facilitated biophysical studies and structurefunction analyses of proteins.

Table 1. Selected uses of recombinant proteins

Purpose of recombinant protein expression	*Examples*
Unlimited amount of biologically active drug	Insulin, growth hormone, erythropoietin
Abundant source of pure enzyme	Restriction endonucleases, Taq polymerase
Protein structural studies	X-ray crystallography, NMR
Structure–function studies	Transcription factors, enzymes
Drug screening	Receptors, enzymes
Raise antibodies	Protein antigen
Detect protein–protein interactions	Yeast two-hybrid assays, farwestern screening

In vitro translated protein

There are now numerous methods and hosts for expressing recombinant proteins, only some of which will be reviewed here. The simplest form of expression involves *in vitro* translation using reticulocyte lysates. In this method, mRNA (usually transcribed *in vitro* from a cDNA cloned into a plasmid) is translated using crude preparations of ribosomes present in reticulocyte lysates. Although the amount of protein is not large, it is possible to label it by including radioactive amino acids (e.g., ^{35}S-methionine). In vitro translated proteins can be used as substrates for enzymes, in catalytic reactions, immunoprecipitations, DNA-protein interaction assays, and so on.

Expression of recombinant proteins in E. Coli

E. coli were an early host for expression and continue to represent an important source for producing recombinant proteins. Expression in *E. coli* works best for relatively small, intracellular proteins that do not require extensive posttranslational modification for function. Proteins are expressed using plasmid expression vectors, many of which can be induced by treatment with reagents such as IPTG (relieves promoter repression), to cause high-level expression before purification.

Attaching proteins of interest to other molecules such as β galactosidase can sometimes stabilize a protein that is otherwise subject to degradation in *E. coli*. It is also common to add a purification tag to the expressed protein to facilitate its recovery after expression. Common tags include glutathione-S-transferase (GST) for isolation on glutathione columns or a polyhistidine tag for isolation on nickel columns.

Advantages of *E. coli* expression include the relative ease of the recombinant DNA manipulations (similar to plasmid cloning) and the rapid selection and expression process. Disadvantages include its inability to perform complex processing like glycosylation, and the fact that some proteins are labile (or toxic) in this host. Also, large proteins are not usually produced or folded efficiently.

Baculovirus-driven expression of recombinant proteins

Baculovirus-driven expression in insect cells is another very common strategy for recombinant protein expression. Using transfer plasmids containing the cDNA of interest, recombination occurs with the baculoviral DNA in Sf9 cells such that the DNA sequence encoding a major viral coat protein, polyhedrin, is replaced with the protein of interest. This results in very high expression of the recombinant protein. Sf9 cells can carry out glycosylation (although structurally distinct from mammalian cells), and it is also possible to coexpress more than one protein subunit (e.g., immunoglobulins).

The recombinant protein often assumes a relatively normal cellular localization (e.g., receptors on the membrane; transcription factors in the nucleus), although the high level of overexpression ultimately distorts this pattern. The high level of expression facilitates purification unless aggregation occurs. However, the requirements of recombination and selection make this system somewhat cumbersome if a large array of mutant proteins need to be investigated.

Expression in mammalian cell lines

For many proteins, it is desirable to perform expression in mammalian cell lines. These systems, while they rarely produce proteins as efficiently as *E. coli* or Baculovirus, have the advantage of processing complex polypeptides using intracellular machinery adapted for this purpose. In addition to requirements for precise peptide processing, folding, or glycosylation, the purpose of producing many recombinant proteins requires their expression in an homologous or tissuespecific system. For example, studies of certain receptors or transcription factors require their expression in cells that contain appropriate cofactors or targets.

In the case of mammalian cell lines, expression of a cDNA is usually driven by a strong viral promoter. Alternatively, inducible expression vectors have been developed that allow selection of a clonal cell line followed by induction of the promoter by the addition (or removal) of reagents (e.g., tetracycline) that modulate promoter activity. Proteins with appropriate leader sequences may be secreted into the media. Although this results in dilution, there are many fewer contaminants in the extracellular media compared to the intracellular contents.

Expression by adenoviral vectors

A variety of viruses have been used for expression in mammalian cells, taking advantage of the natural tropism of viruses and their ability to induce alterations in the intracellular protein synthesis machinery that results in high-level expression. Adenoviral vectors have been of interest because they infect a wide spectrum of mammalian cells and because the adenoviral genome has been successfully altered to create replication-deficient viruses that can harbor mammalian gene sequences.

Adenoviruses have been used to express recombinant proteins in cell lines and tissues that are otherwise difficult to transfect with high efficiency. However, their use as vectors for gene therapy has made them a particularly attractive system for gene expression.

By deleting key adenoviral genes, the modified virus can function as a carrier of mammalian genes. Introduction of the viral vector containing the gene of interest into E1a-expressing 293 cells allows the virus to be packaged and harvested, but it does not replicate in other cells. Adenovirus vectors have been used to express normal genes in defective organs (e.g., expression of CFTR in the lungs of patients with cystic fibrosis), to express toxic genes in tumors (e.g., thymidine kinase in brain tumors), and to target cell-cycle arrest genes in proliferating vascular cells. Limitations of adenovirus include an inability to accept foreign DNA greater than about 10 kb, relatively short-term expression *in vivo* (weeks to months), cytotoxicity, and induction of immune responses. Several other viral delivery systems are under investigation including Adeno-associated virus, Herpes virus, and retroviruses (which integrate in the host genome). Vaccinia viruses have also been used to express recombinant proteins in cell lines, such as HeLa cells, and provide an efficient mechanism for high-level expression of genes in mammalian cell lines.

References

1. *Adams MD, Fields C, Venter JC (1994) Automated DNA Sequencing and Analysis*. New York: Academic.
2. Ausubel FM, Brent R, Kingston RE, et. al. (1995) *Short Protocols in Molecular Biology*, 3rd ed., New York: Wiley.

3. Boultwood J (1997) Gene Regulation: *A Eukaryotic Perspective*, 2nd ed. Totowa, NJ: Humana.
4. Glick BR, Pasternak JJ; *Molecular Biotechnology: Principles and Applications of Recombinant DNA-3rd Edition*, ASM Press, Washington DC.
5. Grompe M, Johnson W, Jameson JL (1998) *Principles of Molecular Medicine*. July 1998, pps. 9-24.
6. Harwood AJ (1996) *Basic DNA and RNA Protocols*. Totowa, NJ: Humana.
7. Innis M, Gelfand DH, Sninsky JJ (1995) *PCR Strategies*. New York: Academic.
8. Klug WS, Cummings MR (2002) *Concepts of Genetics: 7th edition*. Pearson Education.
9. Kneale GG (1994) *DNA-Protein Interactions: Principles and Protocols*. Totowa, NJ: Humana.
10. Landegren U (1996) *Laboratory Protocols for Mutation Detection*. Oxford, UK: Oxford University Press.
11. Latchman DS (1993) *Transcription Factors: A Practical Approach*. Oxford, UK:IRL/Oxford University Press.
12. Leitch AR, Shwarzacher T, Jackson D, Leitch IJ (1994) *In Situ Hybridization*. Oxford, UK: BIOS Scientific.
13. Lewin B (1994) *The extraordinary power of DNA technology. In: Genes V.* Oxford, UK: Oxford University Press, pp. 633-656.
14. Strachan T, Read AP (1999) *Human Molecular Genetics, 2nd edition.* BIOS Scientific Publishers, Oxford.
15. Tijssen P (1993) *Hybridization with Nucleic Acid Probes* (Parts I and II). Amsterdam, North Holland: Elsevier.
16. Trower MK (1996) *In Vitro Mutagenesis Protocols. In: Methods in Molecular Biology*, vol. 57, Totowa, NJ: Humana.
17. Tuan RS (1996) *Recombinant Gene Expression Protocols*. Totowa, NJ: Humana.
18. Vento AB, Gillum D R (2002) *Fact Sheet Describing Recombinant DNA and Elements Utilizing Recombinant DNA Such as Plasmids and Viral Vectors, and the Application of Recombinant DNA Techniques in Molecular Biology.* Office of Environmental Health and Safety, University of New Hampshire (June 3, 2002).

19. Watson JD, Gilman M, Witkowski ZM (1992) *Recombinant DNA*, 2nd ed. New York: WH Freeman.

20. White BA (1992) *PCR Cloning Protocols: From Molecular Cloning to Genetic Engineering*. Totowa, NJ: Humana.

21. Wu R (1995) *Recombinant DNA Methodology II*. New York: Academic.

4

Analysing the Code of Life

In the mid 1980s, genetic technology took a great leap forward with the ability to genotype the DNA itself. The geneticist could now examine the DNA directly without going through the laborious process of developing assays to detect individual differences in proteins and enzymes. Direct DNA analysis had the further advantage of being able to identify alleles in sections of DNA that did not code for polypeptide chains. As a result of these new advances, the number of genetic loci that could be detected increased exponentially and soon led to the identification of the genes for disorders that had remained a mystery for the better part of this century.

Electrophoresis

Electrophoresis is a technique that separates small biological molecules by their molecular weight (Fig. 1). It may be applied to molecules as large as proteins and enzymes as well as to small snippets of DNA and RNA. One begins the procedure by constructing a "gel"-a highly viscous material the actual chemistry of which need not concern us. Purified copies of the biological specimen are then injected into a "starting lane" at one end of the gel.

Finally, a weak electric current is passed through the gel for a specified amount of time.

Gravity and the electric current cause the biological molecules to migrate to the opposite end of the gel. The extent to which any molecule moves depends upon its electrical charge, molecular weight, the viscosity of the gel, the strength of the current, and the amount of time that the current is applied. With constant charge, viscosity, current, and time, smaller molecules will migrate further through the gel than larger molecules.

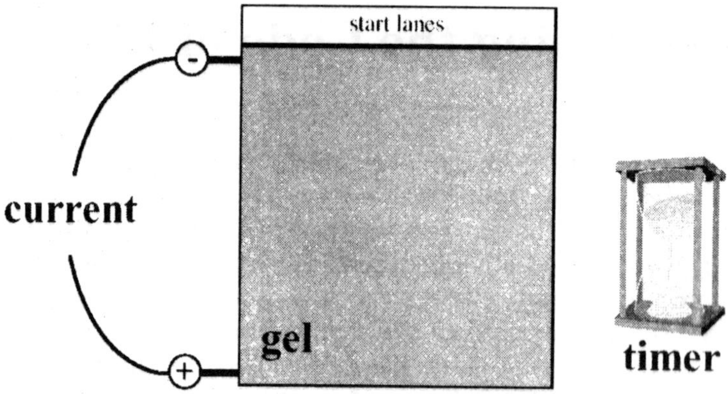

Figure 1. Gel electrophoresis

Cloning

In popular imagination, the term "cloning" is associated with Brave New World, science fiction, and a famous sheep called Dolly. To the molecular geneticist, however, the term "cloning" simply means the copying of a desired section of DNA. Historically, cloning began by isolating a small (i.e., several thousand base pairs) section of human DNA and then, after an elaborate series of steps, incorporating this segment into the DNA of another organism. The "other" organism is called a vector, and of course, not any type of vector will do.

To make many copies of the human DNA segment, one desires a vector that can reproduce rapidly. Hence, the most common vectors are the smallest organisms with dramatic reproductive potential-plasmids, virus, bacteria, and yeast. There are several major reasons for cloning human DNA. One purpose is to obtain large amounts of a human DNA sequence that can then be used as a probe in other types of molecular genetic techniques.

Cloning is also used for therapeutic genetic engineering. Here, the hope is that the vector with a cloned copy of the "good" gene missing in a patient might be incorporated into the patient's cells and produce the missing protein or enzyme.

Probes

A probe is a segment of single stranded DNA or RNA with a known nucleotide sequence. The probe is either synthesized or cloned in a laboratory in very large amounts and is then placed into physical contact with human DNA that has been treated to become single-stranded. Because of complementary base pairing, the probe will bind to the DNA segment that contains the specific nucleotide sequence that complements the probe's sequence. An example is given in Figure 2.

```
                 CTTAAG...                        C
GAATTC...        GACTTC...                        GAATTC...
```

Figure 2. An example of a probe

Light Bulbs

Purified DNA and RNA resemble viscous water. If many small sections of singlestranded DNA are subjected to gel electrophoresis, the sections will migrate to the opposite side of the gel according to their size, but they will be invisible. If probes are added, they will bind to the appropriate complementary snippets of the single-stranded DNA, forming a double strand. But the probes, being composed of nucleotides, will also be invisible.

To locate the probe after it binds to the DNA, it is necessary to engineer the probe so that it carries the biological equivalent of a light bulb. Two major types of light bulbs are used in molecular genetics. The first type is constructed by labeling the probe with radioactive isotopes. The second type of probe uses special fluorescent dyes that allow the probe to be visualized and photographed under specialized lighting conditions. For many types of DNA analyses, radioactive probes are still the method of choice, but fluorescent technology is quickly replacing it.

Restriction Enzyme

A restriction enzyme is an enzyme that recognizes a specific nucleotide sequence and cuts DNA at the sequence. For example, the restriction enzyme EcoRI (for *E. coli* Restriction enzyme number I) recognizes the sequence GAATTC and slices the DNA right after the G. Restriction enzymes are used in a wide variety of techniques. One major advantage is that they can cut DNA (or "digest" DNA, as the microbiologists prefer to call it) into fragments of manageable lengths. Without digestion using restriction enzymes, human DNA segments would simply be too long to allow them to be cloned or subject to many kinds of electrophoresis. Restriction enzymes also play an important role in detecting human polymorphisms.

Polymerase Chain Reaction (PCR)

The polymerase chain reaction or PCR is a technique used to "amplify" DNA-i.e., make a sufficient number of copies of a DNA segment to permit it to be used for other types of techniques. Many people are familiar with the forensic application of PCR. When only a tiny drop of blood, semen, or other biological specimen is available at a crime scene, PCR is used to make a sufficient amount of DNA to permit genotyping.

PCR methodology makes use of many of the concepts outlined above, so it will be explained in some detail (see Figure 3). The procedure begins with purifying DNA from a biological specimen and then heating it almost to the boiling point of water. The heat separates the double-stranded DNA into two single-strands.

Once the DNA has become single-stranded, large amounts of specialized probe are added to the mixture along with an enzyme and a large number of free nucleotides. The probe binds to the DNA and then the enzyme synthesizes a complementary strand to the DNA beginning with the end of the probe and continuing to the end of the DNA segment. If we began with a single copy of DNA, then the end of this process will result in two copies of the DNA molecule (Fig. 3).

By repeating the process of heating, bathing the DNA with probes, and adding the enzyme and nucleotides, we would now have four molecules of the desired DNA segment. Continuous repetition results in a geometric progression of DNA copies-8, then 16, then 32-until a sufficient amount of DNA is available for genotyping. Today, PCR is machine automated and is often done in conjunction with a robot that automatically sequences the amplified DNA. The advantage here is that PCR avoids the complicated laboratory procedures necessary to harvest enough DNA required for other types of genotyping. Automation also reduces the time and cost of genotyping. The biggest disadvantage of PCR is that the technique is so sensitive that it is susceptible to contamination from other DNA. Hence, careful laboratory protocol must be followed.

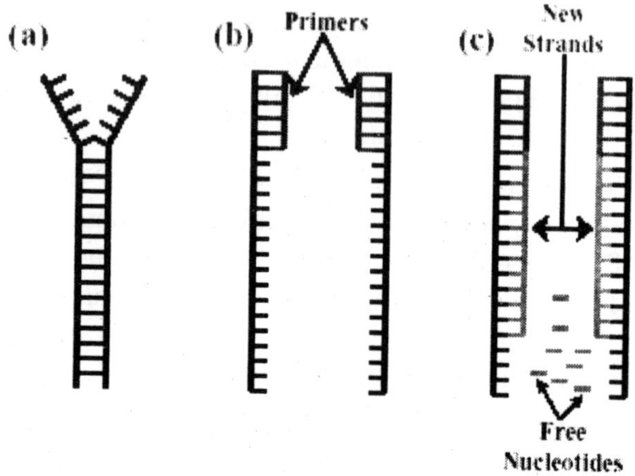

Figure 3. PCR: the polymerase chain reaction. (a) With heating DNA becomes single-stranded; (b) a primer section of DNA is added; it will bind with its complementary base pairs on the original DNA; (c) free nucleotides and enzymes are added that will synthesize a new chain starting at the end of the primer

Types of DNA Polymorphisms

Blood Groups

In the early days of human genetics, the majority of polymorphisms were those associated with proteins in blood. When you blood is typed, you are informed that you are blood group O+ or AB-or A+, etc. The letter in this blood group gives your phenotype at the ABO locus, and the plus (+) or minus (-) sign denotes your phenotype at the Rhesus locus. A number of other loci such as Kell, Duffy, MN, and Kidd can also be phenotyped from blood. These polymorphisms are still used today to assess suitability of donors and recipients for blood transfusions (ABO locus) and to assess Rhesus incompatibility between a mother and her fetus. However, blood group polymorphisms have given way to other, more sophisticated techniques in modern human genetic research.

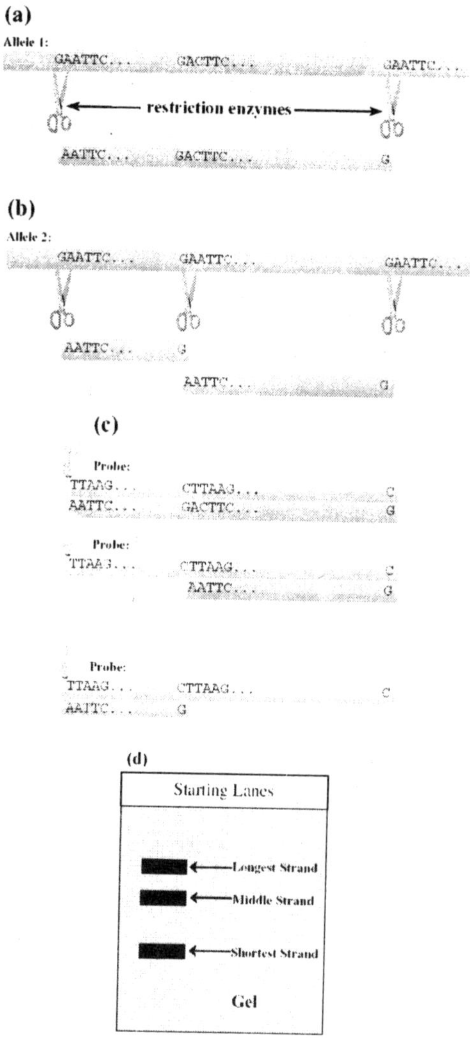

Figure 4. The restriction fragment length polymorphism (RFLP). a): Allele 1 has two restriction sites; incubating this allele in the restriction enzyme will result in one long fragment. b): Allele 2 has three restriction sites; incubating this allele with the restriction enzyme results in two fragments. c): After the restriction enzyme cuts the DNA, the DNA is subjected to electrophoresis that separates the

fragments according to size. A probe is then added to bind to the gene of interest. The binding of the probe after electrophoresis is illustrated in this panel. d): The gel is then viewed under special lighting

Restriction Fragment Length Polymorphism (RFLP)

At this point, it is helpful to describe a genotyping technique that will use all the tools outlined above even though it is not the preferred method for today's research. The restriction fragment length polymorphism or RFLP was the technology that began the dramatic explosion in genetic technology by allowing the field to locate genes for Mendelian disorders like Huntington's Disease and cystic fibrosis. The procedure is illustrated in Figure 4. First a probe must be constructed that will bind with a known sequence of DNA. The sequence is the gene that we want to examine. The probe for this example was previously given in Figure 2.

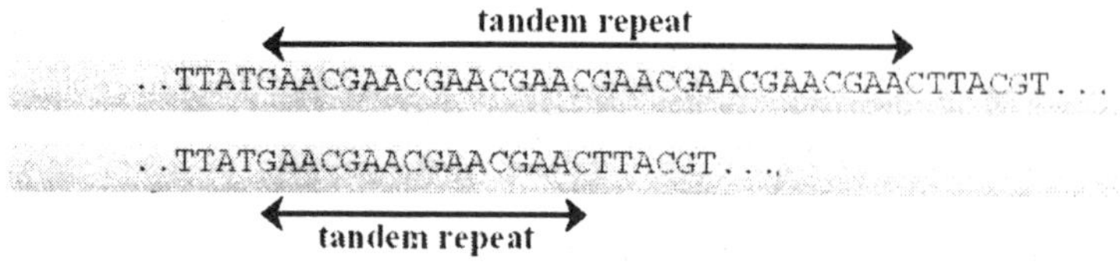

Figure 5. Example of a tandem repeat polymorphism

Tandem Repeat Polymorphisms

Although RFLPs were the first of the modern molecular methods used to detect polymorphisms, they have given ground to other, more discerning techniques. One generic class of polymorphisms has their origin in DNA nucleotide sequences that are repeated a certain number of times, one right after the other. These "tandem repeats" are highly polymorphic in the sense that a large number of alleles may be found at any given locus. For example, one allele at a locus may have the sequence CAG repeated four times, another may have it repeated eight times, while yet a third may have it repeated 20 times (Fig. 5).

Unfortunately, even though the concept of the tandem repeat is quite simple, the terminology for referring to these polymorphisms is quite confusing. For simplicity's sake, we will lump all these fine distinctions together into the single category of tandem repeat polymorphisms. Figure 5 illustrates this type of polymorphism. The repetitive nucleotide sequence is GAAC, which is contained eight times in allele 1 but only four times in allele 2. The probe in this case contains the repetitive complement CTTG.

There are several different ways to genotype for tandem repeats. When the number of nucleotides in a sequence is fairly large (e.g., a series of 25 nucleotides is repeated over and over), then separating alleles by electrophoresis is often done. When the number is small (e.g., repeats of two to four nucleotides), then PCR is commonly used. Here, a special probe for the gene of interest is constructed for the area before the repeated sequence. PCR is then done to amplify the DNA that is then placed into a special sequencing machine that counts the number of repeats.

Single Nucleotide Polymorphisms (SNP)

A single nucleotide polymorphism or SNP is a sequence of DNA on which humans vary by one and only one nucleotide. Because humans by one nucleotide per every several hundred nucleotides, there are likely to be hundreds of thousands of SNPs scattered throughout the human genome. The major advantage of SNPs, however, lies in the fact that they can be detected in a highly automated way using DNA chips. This avoids the laborious process of using gel electrophoresis and thus many more people and many more loci can be genotyped within a given amount of time.

Genotyping begins with PCR to amplify the region containing a SNP. The resulting DNA is then heated to make it single stranded and placed onto a special chip that contains single-stranded DNA and microcircuitry. Specialized computer software can then "read" the specific polymorphism. The potential of SNPs is so great that the Human Genome Project has established a special group, the SNP Consortium, to develop the technology and to locate SNPs.

Gene Sequencing

Nothing surpasses finding the ordering of the nucleotide sequence for a gene, a technique known as sequencing, for the detection of polymorphisms. Although this is the ultimate knowledge for the geneticist and will in time overshadow all other types of polymorphism detection, it is too time consuming and expensive to be used anywhere but in the more advanced research centers.

The Human Genome Project is devoting large amounts of its resources to automating the process of sequencing, so it is difficult to predict the methods of choice by the time these words hit print. Suffice it to say that within the next few decades, the computerized laboratory robots now available to the well-funded genetics laboratory will encounter the dramatic price discounts witnessed by the electronic calculator and later, the personal computer. Every hospital-as well as many laboratories researching individual differences in human behavior-will eventually have a "DNAnalyzer" sitting in the corner of a room spurting out sequence information about genotypes.

In situ Hybridization

In situ (Latin for "on site") hybridization is a technique used with whole chromosomes to: (1) find which chromosome a gene of known nucleotide sequence is located on; or (2) determine if a section of chromosome has been deleted or duplicated. The technique begins by preparing chromosomes much in the same way that is done for a karyotype. The chromosomes are then placed into a solution with a large amount of either radioactively labeled or fluorescent probe.

Combinations of chemicals and heat are used to split the double-stranded DNA in the chromosomes into singlestranded DNA, permitting the probe to bind with the DNA. Excess probe is washed away and the chromosomes are then stained. If a section of a chromosome "lights up" then we know that it has the nucleotide sequence complementary to that of the probe.

Mutational Screening

Once a probe for a section of a protein-coding gene is identified, the curious geneticist can hardly resist the temptation to see if patients with a particular disorder have an irregularity at the locus. A classic example would be schizophrenia and the genes for dopamine receptors. For a long time, it was known that many drugs that diminish the florid hallucinations and delusions of the schizophrenic had important influences on dopamine receptors. Hence, when the gene for a dopamine receptor was first cloned, it was a natural matter to question if schizophrenics differed from nonschizophrenics in this gene.

Except for direct sequencing, many of the techniques outlined above are not suitable for screening the genotypes of patients. For example, RFLPs will only detect DNA differences at a restriction site, and the chances are quite remote that the aberration in the schizophrenic dopamine receptor allele happens to be at just this site. Several other techniques, collectively known as mutational screening, are preferred in this circumstance. The two most widely developed mutational screens use souped up versions of electrophoresis to find small and subtle differences in DNA.

One technique uses a gel that varies in temperature from the top to the bottom of the gel. As the DNA moves through the gel, the increased temperature will cause it to denature into its single strands and the resolution of the procedure permits detection of small sequence differences between a patient and a control. Once such a difference is detected, then the patient's DNA is sequenced to find the exact nucleotide variation(s).

The second technique is conceptually identical with the exception that the gel does not have a temperature gradient. It relies on the fact that small differences in the nucleotide sequence between two strands of DNA can be resolved in different bands on a specialized gel. Again, the patient's DNA is sequenced after a difference is found.

Genetic Influence for Trait

The first task is to make certain that the trait has some genetic influence on it. In the case of simple Mendelian disorders, the rarity of the disorder and the risk to different classes of relatives are usually sufficient to implicate a single gene. More complicated methods must be used for complex phenotypes.

The first challenge is to identify the chromosome on which the gene is located. Without knowing either the gene or its product, it is necessary to use a statistical procedure called linkage analysis to achieve this goal. Linkage begins with a host of genetic loci called markers or marker loci. A marker locus is a polymorphic locus with a known location on a chromosome. Linkage does not try to locate the precise position of the disease gene. Rather, it tries to identify which marker loci are close to the disease gene. In this way, a rough area of a particular chromosome will be targeted for further analysis. With current linkage maps, a disease gene can be located to anywhere within 1Mb (million base pairs) to 2Mb of its exact location. The major advantage of new technology is that it has greatly expanded the number of polymorphic loci that can be used as markers in a linkage study.

Before the human genome was fully sequenced, positive linkage findings were followed up by a procedure called positional cloning or gene walking to identify the location of the disease gene. Because the actual DNA sequence is now known, contemporary researchers power up their computers and download the nucleotide sequence for the region with the positive linkage. They then try to locate genes within this region by looking for promoter regions, initiation codons, stop codons, etc.

As with all human molecular technique, sequencing begins with biological material, almost always blood. Through a series of steps that need not concern us, the DNA from the blood cells is extracted and purified. A very small sample of this DNA is then used for analysis. The physical process of extraction and purification breaks the DNA at large numbers of random places. However, very long fragments of DNA-too long to work with-still remain. Hence, the DNA is often bathed with a restriction enzyme to cut it into fragments of more manageable size. PCR is then used to amplify the region to be sequenced.

The first steps in sequencing are the same as those in PCR (Fig. 6). The DNA is heated so that the double-stranded helix breaks apart into two single strands (a process called denaturing). Then a primer-the identical one used in the PCR-is added. This primer is a small section of single-stranded nucleotides that will bind with its complementary section of DNA that was previously amplified in the PCR.

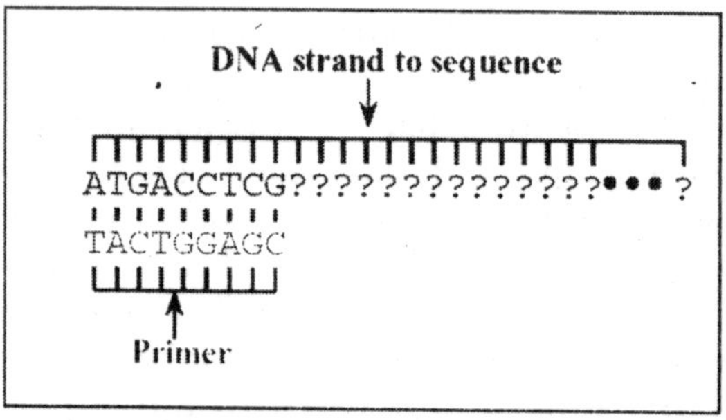

Figure 6. Dideoxy Sequencing. Step 1. Heat the DNA to make it single stranded and add a primer.

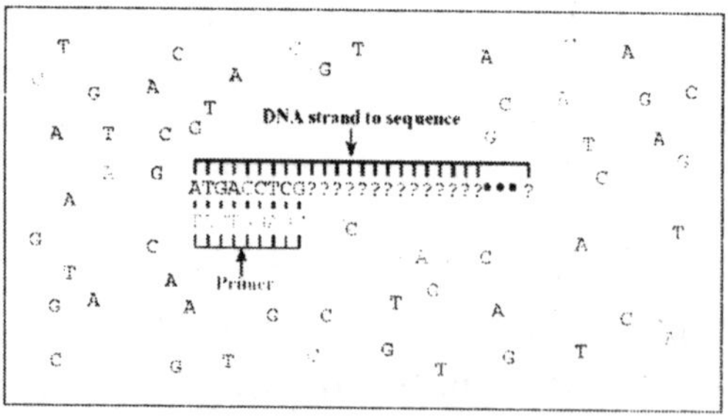

Figure 7. Dideoxy Sequencing.
Step 2. Add nucleotide alphabet soup

The next step is to synthesize the rest of the double-stranded DNA starting with the primer. Essentially, this is the same process as DNA replication but here it is performed in the test tube instead of the cell nucleus. The two major ingredients for replication are: (1) a large number of free nucleotides; and (2) a polymerase enzyme that will build the chain.

Figure 7 shows the DNA with its nucleotide "soup." The majority of nucleotides in the soup have the same chemical composition as the nucleotides in ordinary DNA. The clever trick in this step of the process is the addition of a small amount of specially engineered nucleotides that have two key features. The first is that these nucleotides are chain terminating. That is, whenever one of these special nucleotides gets entered into the chain, the process of building the double helix stops dead in its tracks.

The second important property is that each special nucleotide is "color-coded" through a chemical tag so that it will fluoresce into a specific color when exposed to the appropriate light conditions.

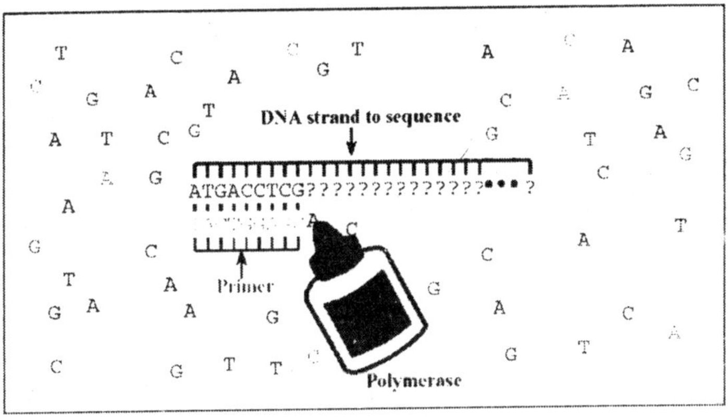

Figure 8. Dideoxy Sequencing, Step 3. Add the polymerase (an enzyme that adds free nucleotides to the primer strand)

Next, a large number of polymerase molecules are added (Figure 8). The polymerase is a complicated enzyme that ordinarily acts to replicate DNA in the cell after the hydrogen bonds are split and the double helix is separated into two single strands. To imbue polymerase with a sentience that it really lacks, one could say that the molecule grabs free nucleotides, inspects the next nucleotide in the single-stranded chain, and then "glues" the appropriate nucleotide partner into the other DNA strand. For example, if the next nucleotide on the single strand is A, then the polymerase will place a T on the growing strand.

The next step is simply to wait and let nature take its course. Because there are millions of single-stranded DNA fragments with primer attached to them, millions of polymerase molecules, and several gazillion nucleotides, many thousands of doublestranded DNA molecules will be synthesized. However, these complementary DNA strands will be of different lengths because of the chance incorporation of special nucleotides.

Whenever one of these special nucleotides is placed into the chain, further synthesis of the DNA strand stops. The net result is many millions of copies of double-stranded DNA, but all of different lengths.

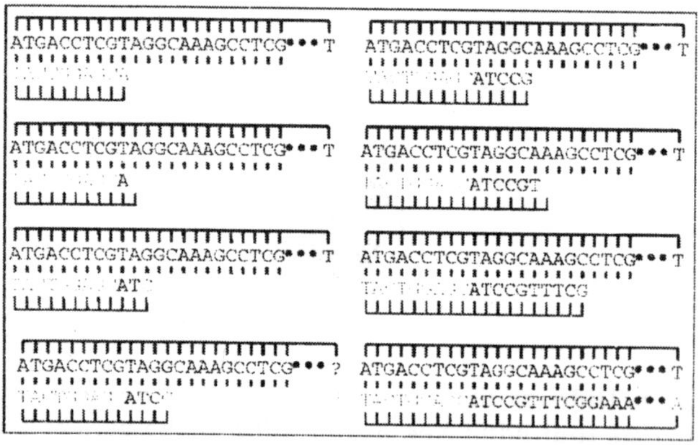

Figure 9. Dideoxy Sequencing, Step 6. Use electrophoresis to separate the strands according to size.

If the DNA mixture is now heated, the double-stranded DNA will break down into its single strands, giving a large number of single-stranded DNA molecules of various lengths. The next step is to load the single-stranded DNA onto an electrophoretic gel. The new forms of electrophoresis are so sensitive that they can detect the difference between two DNA strands that differ in length by only one nucleotide. The result after electrophoresis is completed is depicted in Figure 9, where the fragments would have been loaded on the top of the figure.

When the gel is viewed under the appropriate lighting, the bands will fluoresce. Because the special nucleotides are color-coded, simply reading the sequence of colors gives the nucleotide sequence of the DNA.

This type of sequencing procedure is now highly automated. There are specialized sequencing machines that are essentially computerized robots that perform the automated tasks of timing, heating and cooling, and pipetting mixtures. The newer models also contain a special capillary tube that permits the electrophoresis to be done automatically. Laser lighting and optical scanning allows the colors to be read by a computer instead of a human observer. All the data are processed by specialized software that analyzes the sequence, flags areas of uncertainty, and stores the data.

Many different types of restriction enzymes to cut the DNA. Because one type will cut the DNA at one nucleotide sequence but another will cut the DNA at a different sequence, the resulting sections will have some overlap. By testing which sequences overlap with others, one can reconstruct the whole sequence.

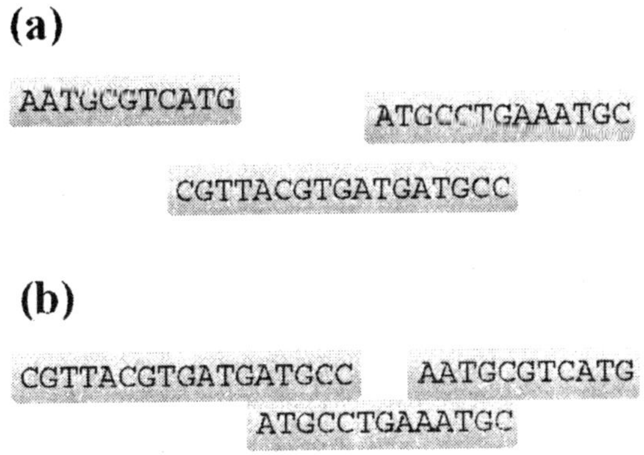

Figure 10. How the human genome was sequenced. (a) Restriction enzymes are used to cut the into a large number of pieces that are then individually sequences; (b) the sequence data for the individual pieces is entered into a computer where powerful software algorithms finds overlapping nucleotide sequences.

Figure 10 illustrates the process of sequencing human genome. Three separate sections of DNA are depicted in panel (a). Two of the sections contain the nucleotide sequence ATGCC, one at the end of its sequence and the other at the beginning of its sequence. Hence these two must overlap. Two of the sections also contain the nucleotide sequence AATGC, again one at the beginning and the other at the end. These two sections must also overlap. The result of placing these sections together gives panel (b) in Figure 13. The problem, then, is analogous to solving a very large, one-dimensional jig saw puzzle.

In reality, the DNA segments are much larger and the number of overlapping nucleotides is greater than those depicted in Figure 13. The DNA nucleotide sequences for the segments are stored in a computer and powerful search algorithms are used to piece them together.

References

1. *Carey G (1999) The New Genetics-Techniques for DNA Analysis.* Human Genetics for the Social Sciences, Department of Psychology & Institute for Behavioral Genetics.

2. Hoban TJ (2004) Public Attitudes Towards Agricultural Biotechnology. ESA Working Paper No. 04-09. FAO, Rome, Italy, May 2004. 16 pp.

3. IFST (2004) IFST Current Hot Topics: Genetic Modification and Food. 32 pp. Institute of Food Science and Technology, UK.

4. IFT (2000) IFT Report on Biotechnology and Foods (4 parts). Food Technol. 54(8): 124-136; 54(9): 53-61; 54(9): 62-74; 54(10): 61-80.

5. James C (2004) Global Status of Commercialized Biotech/GM Crops: 2004. ISAAA Briefs 32. International Service for the Acquisition of Agri-biotech Applications, Ithaca, NY.

6. Metcalfe DD, Astwood JD, Townsend R, Sampson HA, Taylor SL and Fuchs RL (1996) Assessment of the allergenic potential

of foods derived from genetically engineered crop plants. Crit. Rev. Food Sci. Nutr. 36: S165-S186.

7. Nuffield Council on Bioethics (1999) *Genetically Modified Crops: The Ethical and Social Issues*. 177 pp.
8. Nuffield Council on Bioethics (2003) *The Use of Genetically Modified Crops in Developing Countries*. 144 pp.
9. OECD (2000) *GM Food Safety-Facts, Uncertainties and Assessment. Organization for Economic Cooperation and Development*, Paris, France. 49 pp.

5

GM Products and Business of Biotechnology

Food biotechnology has been defined as "the application of biological techniques to food crops, animals and microorganisms to improve the quality, quantity, safety, ease of processing and production economics of food. It thus includes the traditional manufacturing processes used for bread, beer, cheese and various fermented milk products". Genetic modification has been used by humans for at least 10,000 years through selective breeding methods of crops and animals to achieve higher yields, disease resistance and improved quality.

During the past 25-30 years a range of modern genetic modification (or manipulation or engineering) techniques have been developed that enable selected genes to be transferred within or from a species, or from one species to another, using molecular biology techniques including recombinant DNA technologies. While traditional plant and animal breeding required decades for substantial changes to be introduced, modern genetic engineering of plants, animals and microorganisms can now take place over much shorter periods of time, and between species that was previously unattainable. The commercial and societal effects of such technical developments have been substantial, and will continue to be so.

World population has increased from about 4.5 billion in 1980, to about 5.8 billion in 1995 and about 6.5 billion today. Borlaug (2004) and others predict that it may reach 8.5 billion by 2035 and 10 billion by 2050. Food production and availability has also had to increase commensurately. Food crop production increased dramatically during the 20 years of the Green Revolution and in the years beyond. FAO data show wheat yields rose 208% from 1960 to 2000, rice 109%, corn 157%, potato 78% and cassava 36% from increases in yield per hectare rather than increases in area under cultivation, and from increased adoption of modern varieties (up from 9% in 1970 to 63% by 1998), now into second and third generations. Future expansions, however, will need to be provided by land already in use.

With world cereal demand predicted to increase by 50% over the next 20 years driven by increased meat consumption and hence animal feed use, the increased productivity must come from continued genetic improvement of food crops (both conventional and biotechnology) and a reduction in wastage from postharvest losses. Had 1961 average world cereal yields (1531 kg/ha) still prevailed, nearly 850 Million hectares (Mha) of additional land would have been needed to equal the 1999 cereal harvest of 2.06 billion tonnes.

While the proportion of undernourished in the world's population has decreased from 20% in 1990-92 to 17% in 1999-2001 (FAO 2004), and dietary energy supply increased by nearly 0.5% over the same period, most of the world's nearly 800 million undernourished people live in countries with marginal lands and largely depend on agriculture for their livelihoods. The world's poor will need to rely more heavily on biotechnology-related improvements in crop and animal yields and use of marginal lands if the UN's World Food Summit goals of reducing the world's poverty to 50% of 1996 levels by 2015 are to be attained. The principal needs will be further increases in food production (with lack of water as a major constraint), environmental protection, reducing postharvest losses and increasing food quality, both sensory and nutritional. The predicted increase in world population coupled with increasing "affluence" in many countries (as reflected for example in the increasing demand for animal products) will result in the need for twice or three times the amount of food currently available.

Technologies in Genetic Modified (GM) Foods

Genetic changes occur spontaneously over long periods in living cells, but the processes can be accelerated by traditional selective breeding or by modern methods of insertion or removal of genes. Examples of new types of plants developed by hybridisation include triticale and seedless grapes. Modern genetic manipulation enables scientists to cut, copy and reassemble strands of DNA at specific locations in order to insert one or more genes for desirable characteristics, or remove gene(s) for undesirable characteristics.

The techniques to produce recombinant DNA are now relatively routine laboratory procedures with highly specific outcomes. Modern recombinant genetic engineering techniques can be used to transfer genes from microorganisms, plants or animals into cells from each of these living forms. To enable modified cells to be easily recognised in the laboratory, marker genes coding for characteristics such as antibiotic resistance have been included, and this has led to some concerns that these traits may be transferred into microorganisms of public health significance in the human body (e.g. gut), thus increasing their resistance to therapeutic antibiotics. While scientific opinion about the risk of antibiotic resistance marker genes is that it is very low in most cases, their further use is to be discouraged until newer techniques become established.

Plants

The first recombinant DNA-derived food plant was a tomato carrying an anti-sense gene to reduce the level of the softening enzyme polygalacturonase to increase the shelf life of the fruit. Many GM crops have now been developed, including soybean, corn, potato, canola and cotton. In 2003, the estimated global area of GM crops was 67.7 million hectares, of which 41.4 Mha (61%) was herbicide tolerant soybean grown in seven countries, 9.1 Mha *Bt* corn (containing the gene for endotoxin production from the soil organism *Bacillus thuringiensis*) grown in nine countries, 3.6 Mha herbicide tolerant canola (2 countries), and 3.2 Mha each of herbicide-tolerant corn and *Bt*/herbicide-tolerant corn.

In 2003, 30% of the transgenic crop area (20.4 Mha) was in developing countries, up from 14% in 1997. Notably, 90% of the beneficiary farmers were resource-poor farmers from developing countries, whose increased incomes from GM crops contributed to the slight reduction in the proportion of the world's population living in poverty. To date more than 80 crops have been modified for agronomic performance and more than 25000 field trials conducted. The major countries for GM crop cultivation are the USA, Argentina, Canada, Brazil and China.

Animals

The genetic modification of animals can be applied to biomedical research, modelling of human disease, production of proteins or other substances for therapeutic aims, as an alternative source of cell tissues and organs for xenotransplantation, and to improve desired features of farm animals and fish, such as disease resistance and food production. The development of transgenic agricultural animals has lagged behind similar developments for agricultural crops as it is challenging and expensive and because of the low reproductive rate. Cloning of some animals is now possible but raises more issues. GM fish are currently available, but there are no food products derived from GM livestock or poultry. Their role in feeding the world will depend on consumer acceptability.

Microorganisms

Although GM microorganisms and ingredients produced from them (e,g, GM chymosin in cheese making) have been available for many years, use of GM bacteria and yeasts in foods and in food production has been limited. Many applications are possible, including improvement of traditional food fermentations, production of a variety of metabolic end products, and development of bacteria with enhanced probiotic and other health benefit properties.

Advantages of GM Foods

Genetic modification of food raw materials offers several potential advantages and benefits compared to traditional selective breeding techniques. It can:

— provide more food, more economically,

— be faster and cheaper,

— be more precise in selecting particular desirable characteristics,

— allow more traits to be improved, e.g. herbicide and insecticide tolerance, insect resistance, stress (e.g. drought), temperature and virus resistance, salt tolerance, reduce the number of food-deficient regions of the world through the development of stress-resistant varieties more amenable to growing on poor soils,

— improve the shelf life of fresh fruits and vegetables, more easily reduce the levels of allergens, naturally-occurring toxicants and other undesirable constituents in food crops,

— increase the levels of desirable disease-resisting and health-promoting food constituents (i.e. functional foods and nutraceuticals),

— improve the sensory and nutritional qualities of foods (e.g. levels of vitamins A, E, desirable fatty acids, iron, fibre),

— produce a variety of ingredients and processing aids (e.g. enzymes, microorganisms),

— enable optimum feed composition and quality for optimal conversion by animals selected for such feeds, with concomitant benefits for the environment (e.g. development of low-phytate seeds that reduce phosphorus and nitrogen levels in wastes),

— reduce pesticide and herbicide usages, with environmental and cost reduction benefits,

— reduce tillage with consequent soil benefit, and reduce hand-weeding, and

— be used to improve desired features of farmed animals, including fish, such as disease resistance and food production with improved qualities (e.g. lower in fat or fat modified, higher yields of prime cuts, better feed conversion).

Advantages and benefits of GM foods to developing countries are likely to be considerable. With increased populations in the developing world in the decades ahead, coupled with problems of water availability and quality and insufficient land to provide food by conventional means, GM technologies in combination with other technical and socio-economic approaches have the capability to achieve the UN dream of reducing world hunger to 50% of 1996 levels by 2015. Such benefits, however, are not achieved without problems: poor farmers will embrace GM products if it is to their benefit, while for poor non-farming individuals the problem is affording any kind of food, whether GM or not.

There are huge political and economic issues involved that cannot be solved by widespread introduction and consumption of GM foods: they are part of but not the whole solution. Countries must have the necessary infrastructure, financial support (including research expenditure) and expertise to make full use of these technologies, including the necessary regulatory framework to minimise the risks. Many developing countries lack these prerequisites and must be assisted to further build their capacity to meet the challenges for the benefit of their citizens.

Safety Issues of GM Foods

GM foods include the following issues:

— the safety of GM foods, especially in the long term: can they be 'proven' to be safe, and are there

— unintended consequences?

— the environmental impacts of GM organisms and crops,

— the role of "big business" in patenting GM organisms and preventing public access to GM technologies, especially in developing countries, and

— potential allergenicity of novel proteins in GM foods.

GM crops and foods have now been available and consumed for over 10 years, and there appears to be no credible scientific evidence to show that the ingestion of transgenic products is injurious to human health or the environment. GM foods and ingredients currently available on the international market have passed risk assessments conducted by appropriate national and international bodies. Risk is usually defined as "the probability of harm", while a hazard is anything that might conceivably go wrong. Risk is thus a combination of the hazard involved, the probability of its occurrence, and the consequences of that occurrence.

Risk analysis generally follows the Codex Alimentarius principles and consists of the following components:

— Risk assessment is a scientific examination of the nature of the hazard involved (hazard identification), the extent of the hazard and its risk (hazard characterisation), the likelihood of its effects on particular segments of the population (exposure assessment), and an overall assessment of the severity of the risk and its consequences (risk characterisation).

— Risk management is usually undertaken by legislators (state or national jurisdictions) who consider a variety of options before and after their implementation, taking into account, where appropriate, "other · legitimate factors".

— Risk communication involves a dynamic interchange of views among all those involved in assessing and managing risk and the public.

— Risk assessments are thus a science-based examination and interpretation of the facts available in the scientific and technical literature carried out by scientists who are expert in both the topic being assessed and in the modern techniques of risk assessment.

In practice, few traditional foods consumed today have been subjected to any detailed safety or toxicological study, yet they are generally regarded as safe to eat. "Food is considered safe to eat if there is a reasonable certainty that no harm will result from its consumption under anticipated conditions of use". Due to practical difficulties in applying traditional toxicological testing and risk assessment procedures to whole foods, the concept of substantial equivalence was developed. Substantial equivalence is not in itself a safety assessment, but a starting point to identify any intended and unintended differences between the GM organism (plant, animal or microorganism) and its traditional counterpart.

Thus the safety assessment of a GM food compares its characteristics with a conventional counterpart, and involves assessment of factors such as:

- identity
- source
- composition
- effects of processing/cooking
- transformation processes
- the recombinant DNA (e.g. stability of insertion, potential for gene transfer)
- protein expression product of the novel DNA (e.g. effects on function, potential toxicity and allergenicity)
- possible secondary effects from gene expression or disruption of host DNA or metabolic pathways (e.g. macro/micronutrients, anti-nutrients, toxicants, allergens)
- potential intake and dietary impact of the GM food.

Factors are assessed on a case-by-case basis, particularly for GM animals and microorganisms. The precautionary principle states that where scientific evidence for safety is insufficient, inconclusive or uncertain, it should be invoked or be considered as part of the assessment; some have interpreted this as a means of blocking all progress. It should, however, be recognised that scientific evidence can never be conclusive. Science can never prove that anything in life, including food, is safe, i.e. without hazard, because absence of evidence is not evidence of absence.

Environmental Impacts

Much has been made of the potential long term environmental consequences of the dispersal of GM crops and microorganisms into the environment. The issues include their survival in the wild, crosspollination with non-GM crops, transfer of herbicide-resistance to weeds or wild relatives, reductions in insect biodiversity through widespread use of insect-resistant crops, and transfer of selective marker genes (such as antibiotic resistance) in GM microorganisms to gut microorganisms with consequences for antibiotic therapy. To date, after thousands of field trials, there is no evidence of any significant environmental problems with GM organisms. Of concern to organic farmers and those marketing products claiming a "clean, green" image is the potential for contamination of non-GM crops by GM material. Widely reported concerns based on limited experiments that populations of the Monarch butterfly would be adversely affected or eliminated by GM corn have now been shown to be unfounded and the result of inadequate experimental procedures. Pests have not developed resistance to *Bt*, and superweeds have not invaded agricultural or natural ecosystems. There is some evidence of environmental and social benefits emerging, e.g. less pesticide and fuel use, less soil compaction, and less mycotoxin content observed with insect-resistant corn.

Concern has been expressed over potential controls of GM technology through aggressive patenting by some multinational companies, and the inability of subsistence farmers in developing countries to compete, such as the issue of farmers having to pay licensing fees for the modified seeds or not being able to save and use the seed from a previous crop for later planting. However, 30% of areas under GM crop cultivation are in developing countries, and the proportion is increasing, thus farmers in those regions are competing satisfactorily. As noted above, 90% of the beneficiary farmers are resourcepoor farmers from developing countries, whose increased incomes from GM crops contributed to the alleviation of poverty.

Rationalisations amongst large chemical and agricultural companies and poor communications in the past with consumers and regulators in some countries will no doubt lead to changes in marketing strategies and perhaps lead to increased public sector funding of food biotechnology and the capture by public organisations of GM benefits. It is particularly important for research to be directed to indigenous crops of importance in developing countries.

Food Allergens

Food allergens involve abnormal immunological consequences in some consumers to some food constituents, usually particular proteins. The most common types of allergies are mediated by allergenspecific IgE antibodies, and the outcomes can be mild to fatal. The most common allergenic foods (peanuts, soybeans, milk, eggs, some fish, crustaceans, wheat and tree nuts, sometimes called the big 8), account for over 90% of all reactions, although more than 160 foods have been associated with sporadic reactions.

The potential allergenicity of GM foods can be assessed using the decision-tree approach of the International Food Biotechnology Council of the International Life Sciences Institute now adopted and refined by the agricultural biotechnology industry. The assessment examines the source of the gene(s), the amino acid sequence homology of the newly introduced protein(s) to known allergens, the immunoreactivity of the introduced protein with IgE antibodies from individuals known to suffer allergic reactions, and the physicochemical properties (effects of digestion and heat) on the proteins.

The enzyme introduced into glyphosate-tolerant soybeans and the *Bt* proteins used in insectresistant crops have no sequence homology to known allergens and are rapidly digested in mammalian systems. Reports that a Brazil nut gene had caused allergies in consumers eating GM soybean products have been shown to be incorrect, as the transfer of Brazil nut allergenicity to the GM soybean was identified during development work and the product was never released onto the market. Some foods or crops contain natural toxicants (e.g. solanine, enzyme inhibitors), although the major human crops and animals contain few of such constituents through millennia of selective breeding. It is essential that adequate assessments be undertaken to maintain this status with GM foods.

Economic Impacts of GM Crops in Food Sector

The introduction of GM crops has until now mainly addressed the supply side of agricultural crops and food markets. The development of efficiency enhancing GM crops dominates the agricultural applications in most countries where GM crops are grown. In the EU, consumer demand for a continuous supply of agricultural raw materials and processed products at a certain price and a certain quality is seen as the underlying force for the agricultural sector to adapt and to innovate production techniques. Furthermore, the recent reforms of EU Common Agricultural Policy provide several incentives to adapt production quantities to market demand and to put emphasis on quality aspects, both of products and production methods.

Further technological developments and continued increase in GM crop production could affect the future competitiveness of conventional non-GM production. Nevertheless, consumer reaction to GM food has given rise to uncertainty about market developments, in particular the short term prospects for GM products. As a result to consumer concerns, the regulatory framework concerning GMOs has developed and is partially still under review not only in the EU but also in many other countries including the USA.

Labelling has been recommended as a tool to enable consumer choice between products and to avoid further market and trade disruptions. However, labelling systems in which consumers have confidence would require at least segregation of product lines throughout the processing system. Moreover, Identity Preservation would be required to distinguish the different types of products according to their contents of GM material or the way they have been produced whether using GM technologies or not. Segregating and Identity Preservation are attempts to create and establish a separate market for a "new" product, a specific crop. The success of such attempts will depend on supply and demand concerning the new product.

GM crops with enhanced quality traits are most likely to supply niche markets. They are addressing a specific demand and the opportunities for supply are highly dependent on innovations, e.g. new varieties, which provide enhanced quality. On such markets competition and market transparency are generally less advanced than for commodity markets.

Agricultural Trade Systems

Trade of agricultural products today is based on the commodity system. Any system of identification which goes beyond the common specifications would require additional handling effort and would thus create additional costs. Segregation and Identity Preservation are possible responses to consumer demand for specific products.

Commodity System

The bulk commodity system works on the basis that crops from different farms are sufficiently alike to be traded at a common price and to a common grading specification. Usually, commodities from different origins are blended to meet specific grades. For example in the case of wheat protein content, moisture, falling number, specific weight and percentage of extraneous material are taken into account. International trade of agricultural products, in general, is based on the commodity system, which covers about 200 Mio t of grains per year. In the oilseeds sector about 50 Mio t of soya, sunflower and rape seed are traded annually across borders, in addition to 13.5 Mio t of oil from the different seeds and 43 Mio t of meal. Furthermore, many more millions of tonnes are traded on domestic markets under the commodity system.

Bulking up the produce of many producers means that transport and handling costs can be reduced. For example, Cargill has calculated that ocean transport from the US to Europe may only add 13 to the price of a tonne of soybeans if 50,000 tonnes are shipped at a time. The total cost of transportation from a US mid-west farm to European harbours is estimated at 10% of the farm-gate price of soybeans. Furthermore, bulk transport enables a continuous flow for processing, since taking a processing plant down and firing it up again can be time consuming and costly.

Segregation

Segregation refers to a system of crop or raw material management which allows one batch or crop to be separated from another. Segregation is an attempt to create and establish separate markets for differentiated products or to set up a "new" market for a "new" specific product. This corresponds to a dis-aggregation of the supply and demand. Some possible economic effects of market segregation are shown in figures 1 and 2.

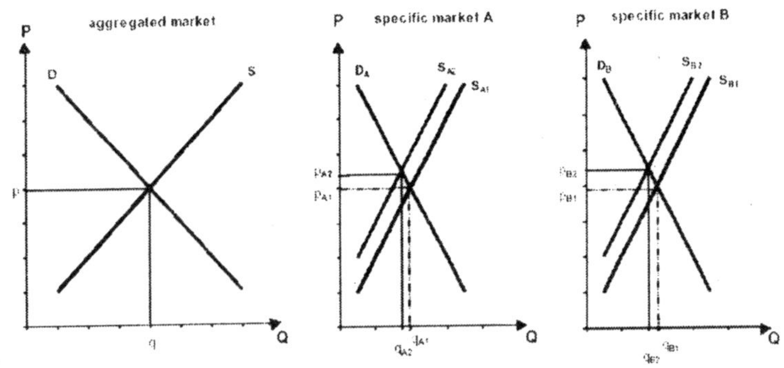

Figure 1. Economic Effects of Market Segregation

In figure 1 it is assumed for simplicity reasons that the aggregated supply for a certain crop would be subdivided equally among the specific markets A and B (dotted lines S_{A1} and S_{B1}). Assuming further that demand would follow the same pattern for both sub-markets ($D_A = D_B$), the price on market A should be the same than on market B ($p_{A1} = p_{B1}$).

However, due to lower quantities produced and traded, potential economies of scale may not be used and production cost per unit might be higher than on the aggregated market. In figure 1 this effect is captured by shifting the supply curves from S_{A1} to S_{A2} and from S_{B1} to S_{B2}. The effect will be a reduction of quantity produced (q_{A2} and q_{B2}) and an increase in prices on both markets (from p_{A1} to p_{A2} and from p_{B1} to p_{B2}). In general, losses in economic welfare can be expected because the potential for trade and specialisation gains will remain partially unused.

Moreover, the assumption of equal pattern of demand on both sub-markets will be unrealistic. More realistic would be the situation as shown in figure 2 with different demand functions on the respective sub-markets (D_{A3} and D_{B3}).

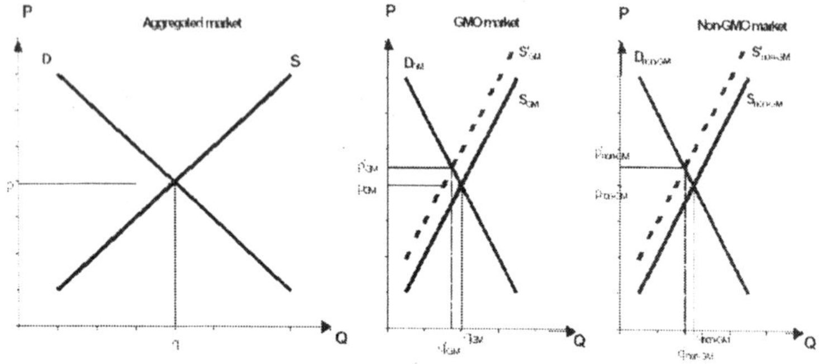

Figure 2. Different demands on segregated markets

In addition, the application of new cost-saving or output enhancing technologies on one of the specific markets would result in a rightward shift of the supply curve. New technologies thus result in price reduction and in higher equilibrium quantity on that specific market. Biotechnology is expected to provide such technological effects, at least in the long-run. Segregation implies that specific crops and products are kept apart, but does not necessarily require traceability along the production chain. In the GMO context, this may pose major problems of liability and consumer confidence. A French investigation identified the absence of labelling requirements at all the stages of a production chain to be the most important difficulty to apply segregation along the production chain and to operate the current labelling requirements for GMOs.

Preservation Systems

Marketing experts have stated that "Identity Preservation programs are the best alternative and the most economical way to meet customer and regulatory requirements". Identity Preservation (IP) is a system of crop management and trade which allows the source and/ or nature of materials to be identified. Thus it goes beyond segregation, since it implies a stronger positive desire to know about the origin of a crop or a product.

The objective of IP is to ensure that a particular crop is monitored throughout the food chain and thus to guarantee certain traits or qualities which might command a premium. IP requires a set of actions to allow traceability and is usually communicated to the consumer by a label. Thus, IP causes additional cost in supply which are illustrated in figure 3.

Introducing Identity Preservation on specific market A would result in a further shift of the supply curve from S_{A2} to S_{A4}. The effect will be a reduction of quantity produced from q_{A3} to q_{A4} and a reduction in price to p_{A4} on market A.

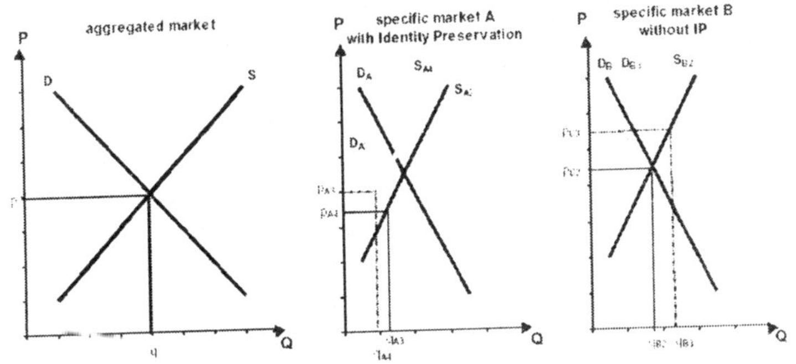

Figure 3. Economic effects of Identity Preservation cost

Currently IP is used to identify crop varieties which provide additional features concerning the content or composition of products (eg, protein content, starch level, oil content). In addition, IP is also applied for features which are not related to the contents but to the method of production (organic food or animal welfare standards) or the geographical origin of the product.

A common example of an IP grown crop under contract is the production of certified seed. Contamination by foreign pollen or other seed varieties has to be avoided and inspections take place to verify purity. The premium for seed wheat production is about 15-20% of the price of a normal wheat crop. This premium should cover the extra work involved for identity preservation.

Other examples for IP systems already in place are related to high erucic acid rapeseed, grown for technical use, waxy corn for starch production and flint corn for breakfast cereals. Identity preservation systems have also been established for certain other specialised (niche) markets, organic produce for example or special varieties of soybeans for tofu production. These products are transported in smaller quantities, reserved trucks or reserved holds of smaller ships.

Compared to the main commodity markets, the quantities currently traded under IP systems are small. Organic food is for instance representing a market share of less than 5%, often less than 1%, in most EU Member States. The highest market shares are obtained for dairy products in Denmark (14.2%) and in Austria (8-10%). In the US, about 100 000 tonnes of soybeans are identity preserved, compared to 75 Mio t harvested under the commodity system.

However, variety choice through IP is seen as contributing more than any other factor to improve the market value of grains. A comparison of recent US prices shows that the premium paid for certain quality traits and for organic products is much higher than the current premium for conventional non-GM crops (s. annex). In the health food sector the price for IP grains and soybeans is about 200-300% of the commodity price.

Identity preservation systems

An important element to establish IP systems is the technical possibility to test samples for the preserved identity (e.g. its physical or chemical contents). Random or regular tests can be carried out for the final product delivered to the consumer or the processor. To enhance the performance, control mechanisms might be applied not only to the final product but also at different stages of production and transportation.

For IP relating to production methods or regional origin, testing of the final product is generally not possible and the consumer has to rely on the integrity of the supplier and the robustness of the IP system. Controls would then have to verify this integrity at different stages of production in order to establish consumer confidence.

Ensuring absolute purity of a food product would be related to prohibitively high costs in practical processing and handling chains. The principle of fixing a tolerance level (threshold) in purity standards is therefore a long-established feature for IP systems throughout the food industry.

Tolerances have for instance been fixed for organic food. Because of the difficulty of eliminating all commingling throughout the production chain, a 5 % tolerance level of non-organic material is allowed in some processed food derived from and labelled as being made from organic ingredients.

The costs of an IP system can be expected to increase with a reduction of the tolerance level. Thus, the setting of a certain tolerance level will be an important factor to determine the costs of an IP system.

Identity preservation often involves advance contracts with farmers who commit themselves to keep the crop separate during harvesting or to produce only under certain rules (quality labels, organic farming). Furthermore, seed varieties, growing specifications, chemical treatments or handling and storage requirements may be subject to specific contracts. With an increasing degree of specification for an agricultural product, which is reflected by a price difference, the likelihood of establishing a contract can be expected to rise.

Identity Preservation and Labelling

There are three approaches for IP related to the introduction of GMOs.

Reasons to consider IP systems in the context of GMOs

The fear of consumers that GMOs could have negative impact on their personal health can be a reason to require traceability. This would allow the identification and if necessary eradication of a harmful modification or product and could be a way to increase confidence in the new technologies. Most crops are living organisms which are able to reproduce a plant. Biosafety considerations require traceability of GM crops to avoid uncontrolled gene transfer and possible danger for biodiversity.

There is a need for processors and traders to meet emerging mandatory GMO-labelling requirements in certain countries, in particular the EU, but also in Switzerland, Australia, New Zealand, Japan etc. The tolerance levels for labelling may differ among countries or still have to be decided. The set of GMOs approved in different countries is not the same. For instance, some corn varieties grown in the US include transformation events not yet approved in the EU. Thus, IP could help to avoid trade disruptions due to differences in the approval status.

Consumer demand for non-GM or GM free food provides an economic incentive for farmers, processors and distributors to supply such products which require IP to be accepted by the consumer. Furthermore, with the development of the second and third generation of GMOs, i.e. specific traits addressing the consumer and the procession industries, IP will become necessary to ensure providing the specific traits to the consumer and to enable a premium for the enhanced value.

EU legislation concerning GMOs, in particular labelling

The release of GMOs in the environment and their placing on the EU market are governed by Directive 90/220. GMOs can only be introduced onto the market after having been assessed and authorised to this end. In February 1998 the Commission tabled a proposal to amend this Directive, which is subject to the co-decision procedure between the European Parliament and the Council. The main objective of this revision is to increase the efficiency and the transparency of the decision-making process whilst ensuring a high level of protection for human health and the environment. With this view, Member States will be required to ensure labelling and traceability of GMOs at all stages of the placing on the market.

Sector-based legislation covers products derived from GMOs, in particular GM food. This legislation has to be further specified and extended, in line with the revision of Directive 90/220. The Commission will table a proposal on Novel Feed, including GM feedstuffs, in the second half of 2000. The White Paper on Food Safety identified a number of actions to re-establish public confidence, in particular completing and harmonising labelling requirements. The Novel Food Regulation and the seed legislation already provide for mandatory labelling of food and seeds containing or consisting of GMOs. Two GM varieties, one of corn and one of soya, and their derived products were already on the EU market before the Novel Food Regulation came into force. Therefore these two varieties were not subject to additional labelling requirements. In order to ensure the labelling of these varieties it was necessary to ensure the appropriate labelling through a specific regulation.

With the adoption of this labelling regulation, the Council invited the Commission to study the practicability of setting down a de minimis threshold which takes account of the problem of adventitious contamination. In response to this request the Commission has adopted a regulation which fixes a tolerance level of 1% for each single ingredient on the condition that the operator has taken appropriate steps to avoid the use of GMOs as a source. A (non-exhaustive) list of food ingredients or foods comprising a single ingredient in which neither protein nor DNA resulting from GM is present, shall be drawn up.

The negative list is a concept applicable to processed GM products in which no genetically modified material can be detected any more. These products would be exempted from compulsory labelling. In January 2000 the Commission also introduced labelling requirements for additives and flavourings that have been genetically modified or have been produced from GMOs. The Commission White Paper on Food Safety proposes the harmonisation of labelling rules for food, additives, flavourings, clarification of the authorisation procedures in the Novel Food Regulation and the establishment of a legislation concerning food and food ingredients produced without genetic engineering.

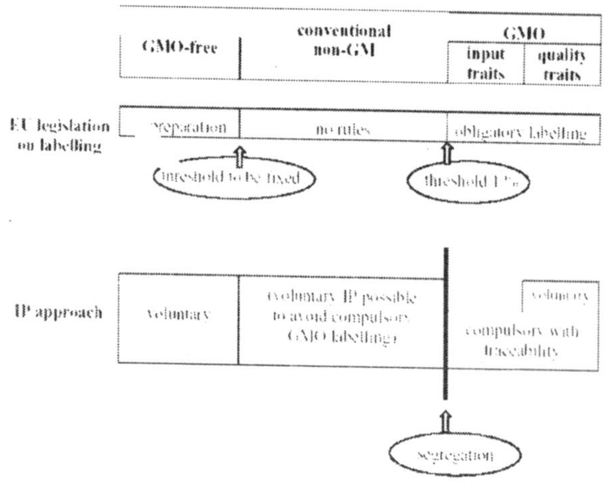

Figure 4. Labelling and Identity Preservation

Following the current EU legislation on labelling and the general features of Identity Preservation systems, three different approaches to IP have been identified in the GMO context (Figure 4).

IP systems are common practice for crops that have a specific value to their consumer, for example through improvements in nutritional value, colour, texture, flavour or processing properties. With the development of new traits by biotechnology, the economic incentive for IP would increase. In addition to the labelling requirements under the novel food regulation, there would be a clear incentive on the supply side (farmers, processors and retailers) to introduce IP and thus to preserve the additional value or quality of such a GM crop through the processing chain. IP would distinguish a product for which consumers are expected to pay more than for a conventional product.

The economic viability of a GM product with consumer oriented traits will depend largely on the ability to identify these specific values in a cost efficient way. In general, detection and quantification of modified DNA and protein depend on the availability of appropriate reference material.

GMOs offering specific qualities to the producer and the final consumer will only be accepted if these qualities can be guaranteed within a certain tolerance level. Tolerances will have to be fixed in accordance with the purity expectations of the buyers of these products.

The second approach for IP is to preserve and label GM-free products in order to enhance consumer choice. Current EU legislation already requires compulsory labelling for food containing GMOs. Thus, the introduction of labelled GMO-free food would in theory enable the choice between three categories of foodstuffs: novel GM food; conventional non-GM food and GMO-free products.

Some European food trade companies which are trying to serve the non-GMO market niche are claiming that they are supplying non-GMO products. However, the explanations and guarantees given to the consumer sometimes lack a sufficient transparency. However, it can be expected that the share of conventional food will diminish over time, since the pay off for GMO-free products can be expected to be higher than for conventional non-GM products. If producers decide to make an effort to segregate, the additional costs to comply with GMO-free standards might be low compared to the additional premia achieved on the market. On the other hand, if at least part of the consumers accept labelled GM food, some conventional raw material would enter into GM-labelled final products.

Testing: Reliable testing will be necessary to prove that a product does not contain or contains only a limited percentage of GM material. A workshop on GMO research perspectives held in the context of the EU Fifth Framework Programme identified the "development of rapid, reliable detection methods for GM foods and their derivatives" as one of the top priorities for further research in the GMO context.

For international trade a standardised test would help to avoid liability conflicts and trade distortion in case of different labelling requirements and approval status. The American National Grain and Feed Association, for instance, has called for the introduction of an accurate, repeatable and low cost test to distinguish between conventional and GMO products. USDA has recently announced that it will establish a reference laboratory to evaluate the validity of analytical procedures and to establish sampling procedures for use in testing bioengineered grains and oilseeds.

The more expensive testing for GMO, the more likely will be arrangements based on declaration of honour concerning the GMO-free status of a product. This would imply a certain system of field and production control to satisfy consumer confidence.

Tolerance: Most crops can easily be contaminated with other material by pollen drift or by mechanical commingling during harvest, storage or transportation. The debate about tolerance levels has raised other questions concerning the non-GM status of a plot or farm on which GM crops have been grown. The argument is that inherited modified genetic sequences are likely to persist on the farm-for example in the case of rape-even after the crop has been harvested and sold.

Currently most US extension services have warned farmers to be careful when signing a contract to supply non-GM or GMO-free products, since accidental contamination cannot be excluded. Trading GM crops as part of a commodity system would result in losing their track within the transportation and processing chain. Thus any commodity sample originating from a region or country where GM and conventional crops are grown in parallel might contain GM crops. Traceability, i.e. a compulsory IP system, has been introduced as a strategy to re-establish consumer confidence in the EU beef sector following the BSE crisis. Traceability could also be a strategy to monitor the environmental and health effects of GMOs and to enable choice to those consumers who want to avoid GMO consumption.

Trade Issues

While accounting for the main producer of GM corn and soybeans, the US are the leading exporter for these commodities. Argentina is the second biggest producer of GM soybeans and the third exporter. The main importing countries for these commodities, the EU and some South-East Asia countries, have taken a restrictive stance on GM food. In particular, labelling of the GM nature of food ingredients is compulsory in the EU. Japan intends to implement mandatory labelling by the second half of 2000.

This situation has become a trade issue. However, it is difficult to isolate the possible effect of biotechnology on developments in trade, as many other factors play a role, like changes in competitiveness, transportation costs and the transaction costs of giving up of long-established trade links. The issues at stake are of a different order of magnitude for soybeans and for corn. Between 1995 and 1997, EU imports from the US were worth, on average, 2 billion Euros for soybeans and soymeals and 0.03 billion for corn. In addition, EU imports of Corn Gluten Feed are estimated to be worth around 500 Mio.

US soybean exports declined from 26 to 20 Mio tonnes between 1997 and 1998, while world soybean trade held fairly steady. EU soya imports from the US have been partially replaced by imports from Argentina. The USDA has concluded that "traditional competitive forces (primarily prices) appear to be the main driving factors behind the changes in observed bilateral trade patterns". As the share of GM soybeans is much higher in Argentina than in the US, this shift in trading pattern cannot be attributed to reluctance to import GM soybeans.

The drop is even sharper for corn than for soybeans. US corn exports fell from 60 Mio tonnes in 1995 to 41 Mio in 1998. Most of the drop occurred on South-East Asia markets (with the exception of Japan) and is explained by the situation of China, which became again a net exporter of corn. On the EU market for corn, the share of US has steadily fallen while the share of other partners, in particular Argentina and Hungary, has significantly increased. The USDA considers that the loss of shares on the EU market results from issues related to biotechnology, in particular the differences in regulatory approaches.

While 11 types of GM corn have been approved in the US, only 4 have been cleared at EU level (table 1), and some Member States have decided to suspend authorisations for growing. Non-authorised GM crops cannot be placed on the EU market. In the absence of tolerance thresholds, if traces of such crops are found in a given consignment, it cannot be cleared for importing into the EU. According to the USDA, this situation has created uncertainties.

However, the type of GM soybeans which is mostly grown in the US (herbicide tolerant) is authorised in the EU for imports and processing (but not for growing purposes). According to the USDA, only a small part of US areas have been sown to non-EU approved corn varieties and the EU only accounts for 1% of US corn exports.

Table 1. Approvals of GM crops in the EU and the US

	US		*EU*		
GM crops	*approved*	*% sowings*	*approved*	*pending*	
Corn	11	35%	4	5 among which	1 already approved for imports & process 2 are the same GM crop but with different uses
Soybeans	3	50%	1	none	
Rapeseed	3	15%	4	3	only one is same as in US

Trade issues have been addressed in the Biosafety Protocol, which aims at ensuring an adequate level of protection for transfer, handling and use of GMOs which might have an adverse effect on biodiversity. Reference is made to the precautionary principle in this respect. It is hoped that procedures foreseen under this Protocol, in particular information sharing and accompanying documentation, will help improving the predictability of transboundary movements of GMOs (Table 1).

Biosafety Protocol

The Biosafety Protocol provides a framework for addressing environmental impacts of bioengineered products that cross international borders. It was concluded in Montreal in January 2000 by delegates from 138 countries. In accordance with the precautionary approach, the objective of this Biosafety Protocol is to contribute to ensuring an adequate level of protection in the field of safe transfer, handling and use of Living Modified Organisms resulting from modern biotechnology that may have an adverse effect on the conservation and sustainable use of biological diversity, taking also into account risks to human health, and specifically focussing on transboundary movements". The procedures foreseen under the Protocol are different for Living Modified (LM) seeds and commodities.

Advance Informed Agreement procedures shall apply before the first transboundary movement of seeds. Notification of exporter before movement. Accompanying documentation with precise identification and requirements. For LM commodities used as food, feed or for processing:

— Information sharing on approved LMOs through Biosafety Clearing House. Possibility for developing countries without domestic regulation on LMOs to take decisions on imports under the Protocol, to benefit from assistance (financial, technical, capacity-building).

— Documents accompanying transboundary movements of LMO commodities stating that they "may contain LMOs".

In addition, as already mentioned, the EU regulatory framework is under revision. Changes are also considered in the US and in many other countries. Identity Preserved markets are expected to increase in number and market share, with or without GMOs entering the markets. Trade experts have estimated a 25% market share for IP corn and IP soybeans by 2005.

References

1. *Borlaug NE (2004) Feeding 10 billion people-our twenty-first century challenge*. Special contribution 1. The State of Food and Agriculture 2003-2004, FAO, Rome, Chapter 3, pp 2-3.
2. CBD (2000) *Cartagena Protocol on Biosafety to the Convention on Biological Diversity*. United Nations Environment Programme. January 2000. 24 pp.
3. FAO/WHO (2000) *Safety Aspects of Genetically Modified Foods of Plant Origin*. Report on a Joint FAO/WHO Expert Consultation on Foods Derived from Biotechnology, 29 May-2 June 2000, Geneva, Switzerland. 35 pp.
4. FAO/WHO (2003a) Principles for the Risk Analysis of Foods Derived from Modern Biotechnology. CAC/GL 44-2003. Rome, Italy. 2003. 4 pp.
5. IUFoST Scientific Information Bulletin on Biotechnology and Food, International Union of Food Science and Technology, February 2005.
6. FAO/WHO (2003b) Guideline for the Conduct of Food Safety Assessment of Foods Derived from Recombinant-DNA Plants. CAC/GL 45-2003. Rome, Italy. 16 pp.
7. FAO/WHO (2003c) *Guideline for the Conduct of Food Safety Assessment of Foods Produced Using Recombinant-DNA Microorganisms*. CAC/GL 46-2003. Rome, Italy. 16 pp.
8. Hefle SL, Nordlee JA, Taylor SL (1996) *Allergenic foods*. Crit. Rev. Food Sci. Nutr. 36: S69-S89.
9. Hess, CE (1992) *Biotechnology-derived foods from animals*, v.32(2): p.147-150.

6

Big Money and Agri-biotechnology

The first successful breakthroughs in agri-biotechnology were achieved in the mid-nineteen eighties (Table 1). Small and medium sized enterprises (SME), acting as start-ups, have had an active role in initial developments, both in the United States and in Europe. Due to financial and economic crisis in European Union (EU) in early nineties, these SMEs had experienced difficulties in finding sources of financing. Many of them have been brought up by the emerging "life sciences industries". A typical example is Plant Genetic Systems (PGS), a European small enterprise, which succeeded in 1985 to integrate *Bt* genetic material into a tobacco plant. In 1996, AgrEvo, the newly formed merger between Hoechst and Schering bought PGS.

Life sciences industries had experience in biotechnology, but for most part, this was limited to pharmaceutical applications. They were, however, able to transfer this knowledge to the area of agri-biotechnology and/or to take on board the experience of the start-up SMEs. In addition to creating common platforms for research and development, they were in a position to implement and amplify agri-biotechnology, on the basis of their plant protection activity. While pharmaceutical markets are more or less narrow, the markets for agro-chemicals are wider and trans-national, even if authorisation procedures are somewhat different from one country or group of countries to another. Experience of the life sciences industries in authorisation procedures, in patenting and also in introduction to the market has been another key factor for the extension of their biotech activities. However, authorisation procedures for medicines, pesticides and GM crops are very different.

Table 1. Stages in development of the agri-biotech industry

Years	*Stage*	*Developments*
1983-1994	Exploratory	Spin-off, from Universities to SME's start-up SME's late 80/early 90: economic and financial difficulties.
1994-1998	Consolidation	Emerging life sciences trans-national companies progressively buy: a) biotech SME's b) seed companies.
1998-?	Adding value	Biotech industry seeking new agreements: a) with research/development partners b) with food processors

Patenting and Origin of Agri-biotechnology

Patenting of agri-biotechnology and the breakdown between private and public research have become new issues. Like previous agricultural innovations, biotechnology is industry-driven, however, biotechnology uses and modifies living organisms. It has been developed by life science industries, while impetus in conventional breeding had been initiated by public research. This translates into new patenting rules on plants: patents on GMOs are very different from traditional "obtainers' rights". Crop varieties obtained through conventional breeding can only be produced and sold by their "obtainers", but other seed companies can use these crops for further genetic improvement (breeders' exemption).

Patents on biotechnology were first introduced in the US, following a 1980 ruling of the Supreme Court, which authorised patents on living organisms. According to Joly, in 1993, US firms owned 70% of the world patents on plants. Biotech patents apply to the "genetic event", that is to say, to the transformation which has been introduced in the plant. They protect the private origin of the technology. In some cases, technological fees charged by biotech companies when selling their genes or GM-seeds allow for remunerating the new technology or trait. The legal framework applying to patenting of biotechnology within the EU is complex.

The European Parliament and the Council adopted in June 1998 a Directive addressing the legal protection of biotechnology. This Directive foresees several exemptions, in particular for human body, plant varieties and animal breeding. It also takes over farmers' exemption: farmers are allowed to save GM seeds for the purpose of own production. In principle, no patent can be obtained for plant varieties, only for inventions (transformation event) relating to them. Plant varieties are covered by a specific regulation which includes breeders and farmers' exemption. However, the European Patent Office has recently indicated that patents on plants and animal are not excluded by the European Patent Convention.

Firms considered that patents were not protective enough, because of possible overlaps. For example, there are many types of *Bt* transformation events. As a result, in addition to patents, registered trade names are usually used to further protect and identify the new technology. Quite often trade names of GM seeds refer to corresponding agro-chemicals, in particular for herbicide tolerant crops. Typical examples are Liberty Link and Round Up Ready crops. Patents on herbicide tolerance are often seen as a way to prolong the effect of expired patents on herbicides. These examples of combined products illustrate the importance of another synergy between biotech and crop protection activities.

Many biotech firms are selling both the GM technology/seed and the agro-chemical product to which it is combined. It allows for "technology-package" or combined marketing, including adjusting prices of both products and using the existing distribution and consulting channels for crop protection products. Beginning in the last quarter of 1995 up to the first half of 1999 the life sciences sector has been characterised by a large number of mergers, acquisitions and joint ventures. Four of the important factors that are driving this consolidation process are:

— The development of new genetic traits that are able to (1) increase the efficiency of farm production; (2) offer new product specifications for industrial or end users.

— Synergies, whereby research capabilities and technology are shared across multiple product lines.

— Closely linked to the above point are economies of scale in research and development in the area of agrigenomics, marketing and a whole host of other functions. Such economies of scale are of strategic importance, considering the need to invest vast sums of money in regard to biotechnology to develop new GM traits.

— Intellectual property rights create barriers to entry.

Extending and securing access to the seed market has been a driving force for a second wave of acquisitions and agreements, resulting in a further consolidation within the agri-biotech sector. Concentration has diffused from the agro-chemical sector to the seed sector as key "life science" players have become leaders in both fields.

While carrying out further research on second-generation GM plants, which will include quality traits, biotech firms have adopted a commercial strategy for preparing the introduction of these new generation plants. On their upstream side, they have entered new agreements with genomics companies, to increase their research/technology portfolio.

On the downstream side, biotech firms seek to invest further down in the food chain. They have concluded or are considering agreements with food processors. The food industry is in a key position. In Europe, many food processors are trying to avoid GM ingredients because of consumer reluctance.

Significant changes in the biotech sector have occurred in 1999. Growing consumer concern, extended public debate and food suppliers' initiatives have had a feed-back effect on the biotech industry. There has been a gradual slowdown in mergers and a shift towards joint ventures or agreements with genomic research firms or institutes. Some biotech companies have also offered to make their technology or experience available to public research centres, in particular for developing countries (several examples with GM rice). By doing so, these firms are trying to respond to public criticism and to improve their strategy and image. Another significant change is the separation of the pharma and agri-biotech activities of some leading biotech companies. This might represent the start of a new phase in restructuring the agri-biotechnology sector, the move away from a "global life sciences" strategy.

Crop Protection

Generally, the share of biotechnology in the agri-business part of life sciences industries is not indicated in financial reports or in publications. The share of agriculture in life sciences activities provides a first identification of agro-chemical and agri-biotech companies, as indicated in table 2.

Table 2. Importance of agriculture within "life sciences" business

Company	*Estimated % of total sales for the agri-business section, 1998*
AgrEvo	100%
Monsanto	47%
Novartis	26%
Rhône-Poulenc	19%
Astra-Zeneca	18%
DuPont	13%
Dow Chemical (agri subsidiary is Dow Agrosciences)	9%

Seed Market

In the early nineties, the concentration rate in the seed sector was not as high as for agro-chemicals. In 1994, the Top 12 of the seed industry accounted for 20% of the world market in seeds. Rural Advancement Foundation International (RAFI), USA estimates that the top 3 now controls 20% of the market. The commercialisation of GM seeds that began in 1996 resulted in a series of mergers, acquisitions and alliances between gene providers and some of the major seed companies. Many of the major agri-biotechnology companies now have access to the seed market in which they can market their biotech products.

Biotech firms have adopted different strategies with regard to commercialisation of seeds. Some are mainly commercialising their GM-products via their seed subsidiaries. Others have concluded agreements with seed companies. In the framework of such agreements and generally against payment of a technological fee, biotech companies are selling their GM-products to seed companies. The latter integrate the genes of interest in the germplasm of their leader varieties, via conventional breeding or back-crosses. This allows for further improvements in the performance of GM-crops. The availability of germplasms with characteristics which are adapted to local conditions or demand is also a key factor. In this respect, breeders as well as small and medium-sized seed companies still have a role to play. For this reason, many agreements concerning seeds are implemented on a regional/national basis. While there is a global consolidation in the seed sector, there is still a regional/national dimension.

The case of Limagrain, a French group which ranks among the world top-5 of seed companies, is illustrative. While being a world leader in seeds, this group has kept a strong regional implantation. It has developed on the basis of a 50 year-old farmers' co-operative and is independent from the main life-sciences players. It holds various subsidiaries. Limagrain Genetics international is active in seed business in Europe and Northern America for arable crops and another subsidiary (Vilmorin) is a leader for vegetable seeds. Other subsidiaries of Limagrain produce food ingredients and bread. Limagrain has developed biotechnology as a mean to meet processor and consumer specific needs, in the framework of organised supply chains. In response to demand for non-GM seeds and crops in Europe and in Japan, Limagrain has also organised non-GM maize supply chains. Limagrain sells GM seeds in Northern America, but not yet in Europe.

Biotechnology has increased concentration in the seed sector, as if the existing concentration in the agro-chemical sector had spread to the seed sector. At the beginning of 1999, according to RAFI, the Top 10 of the seed industry controlled around 30% of the world market. Just four companies (DuPont/Pioneer, Monsanto, Novartis and Dow) controlled at least 69% of the North American corn seed market. The same four companies control 47% of the American soybean seed market. As a result of this consolidation process driven by biotechnology, there is an increased concentration on the input side of the farming sector. Life sciences companies form an oligopoly for supplying inputs for crop production, furthermore after having completed mergers or agreements with seed companies.

Marketing Strategies

The marketing strategy developed by biotech firms has been focused on farmers, the first customers interested in agronomic traits of GM crops. They have shaped farmers profitability expectations. The importance of agriculture within the life sciences business provides a first view of the position of the agri-biotech firms. Further elements have been taken into account to establish profiles and to assess the strategy of each of them. In particular, they have been considered from the upstream/downstream perspective with regard to agriculture. The aim is to assess whether firms had an input-or an output-oriented strategy. For example, if a company has built on the basis of an experience in crop protection and has heavily invested in the seed market, it is considered to be more input-oriented. Farmers are key customers for these firms. Conversely, if a firm has already developed quality-traits crops and has substantially invested downwards in the food chain, it is seen as output-oriented. Food processors are key customers, but farmers remain partners for growing the crops.

About profiles of the main biotech firms. The profiles of the main biotech-players have been established on the basis of key financial indicators (e.g. sales, acquisitions etc.) and factual information (e.g. products) available in their annual reports, on their internet sites, or in various press releases. There is no clear-cut border between input-oriented and output-oriented strategies. In recent years, most of the biotech firms have implemented an input-oriented strategy, focused on the farm sector rather than on the food-processing sector. In their global life sciences strategy, they have failed anticipating differences in perceptions of science in various countries. They have overlooked the question of consumer acceptance and have underestimated the reactions of retailers and food processors. Market structures and organisation play a significant role. Retailers have a strong market power, in particular in Europe. The food processing industry is concentrated, in particular in the US.

The downstream side of the food chain will be of key importance for the next generation of GM crops. Many biotech firms already have quality-traits crops in their pipeline, and rely in the long run on an output-oriented strategy. Some quality-traits crops are already on the market. They are of interest for the feed and livestock industries as well as for the non-food sector. The US meat industry is already concentrated. Biotech is inducing further integration on the upstream side of this industry, acting as a possible vector between the crop and the livestock sector, via GM feedstuffs or animal health products like vaccines.

The emerging clusters of firms that control the food system from gene to supermarket shelf. Following are the clusters:

— Cargill/Monsanto. Cargill did not have access to GM seeds, thus it formed a joint-venture with Monsanto. In addition, Cargill enters a merger with Continental grain, one of the 4 major grain elevators in the US (together with Cargill, ADM and Bunge). This merger will bring them in a key position for exports, hence it is also illustrative of globalisation. Together, these three firms own a complete food cluster.

— Another cluster is the Novartis/ADM connection. It was established through the joint-venture between Novartis and Land O'Lakes to develop speciality corn hybrids.

— Conagra provides the emblematic example of wide ranging clusters, "with diversified interests ranging from farm gate to dinner plate". Conagra is a leader in the US for grain milling, feeding stuffs, slaughtering and meat processing. One of its subsidiary, United Agri Products, is a leading distributor of crop production inputs: fertilisers, crop protection and seeds, including GM ones.

— A similar case is DuPont, a trans-national firm for which the activity spectrum is even wider, from "dirt to dinner plate", dirt referring to the chemical business of DuPont.

Biotech has already generated increased concentration on the input side of the crop sector. Considering the emerging "gene to supermarket" clusters, biotech finally appears as a driving force for vertical integration and for further consolidation throughout the agri-food sector, from the upstream to the downstream side. The position of farmers in this rapidly consolidating context is an issue of concern.

With the development of biotech, farmers more and more depend on a limited number of suppliers for crop protection. In addition, farmers adopting biotechnology are confronted with several constraints. GM seeds are often sold in the framework of contracts which generally preclude seed-saving by farmers. Some biotech companies have taken action against producers who attempt to save seeds, on the basis of "infringement of intellectual property rights". Saving seeds for further sowings is a long-established tradition, at least for crops allowing for doing so (e.g. wheat, soybeans-25% of soybean seeds are estimated to be farm-saved).

Biotech firms have developed technologies that render GM crops sterile. The initial name, "Technology Protection System" has been turned into "terminator technology" by NGOs that have denounced the costs for farmers and the loss of independence. Faced with overall criticism, Monsanto announced in late 1999 that it would not implement the terminator technology. The Technology Protection/Terminator System had also been supported by public research. Various reasons for justifying sterility of GM crops have been put forward, in particular protecting the research-value of GM seeds and limiting gene-flow into the environment.

When selling their technology, some biotech companies are charging a "technological fee". The technological fee is presented as a coverage for research costs and allows for a margin of profitability for biotech firms. It results from the private origin of the new technology and has to be considered together with property and patenting rights. Generally, the technological fee is first paid by seed firms (which are sometimes subsidiaries of biotech companies), and is later transferred to farmers. The technological fee and the restriction on seed-saving imply increased seed costs-as such costs are to be paid each year-and a loss of autonomy for farmers.

As far as HT soybeans are concerned, the American Soybean Association (ASA) has recently complained about significant differences in prices of Round Up Ready soybean seeds between the US and Argentinean markets. According to ASA, a bag of such seeds costs 12 US $ more in the US, and part of this difference is attributable to the 6 US $ technological fee, which is apparently not charged in Argentina. The combined pricing strategy and the observed variations in GM seed prices point to the existence of margin of manoeuvre for biotech firms. The market power of seed and agro-chemical suppliers deserves further assessment.

Biotechnology has generated increased concentration on the input side of the crop sector. This raises the question of increased dependency of farmers on a limited number of suppliers for crop production. Moreover, some biotech firms have already concluded agreements with grain processors, as is the case with the Monsanto/Cargill cluster. The downstream side of the food chain is also quite concentrated, either at the level of food processors or at the retailing industry. Biotech is very likely to be a driving force in such a process, for two reasons.

— GM seeds are often sold and sown under contract. GM crops require adjustments in growing and management practices.

— If segregation or identity-preservation develop, crops, be they GM or not, will increasingly be grown and sold in the context of contracts.

Adoption of GM Crops

The adoption of GM crops by farmers in the US, Canada and in Argentina has proceeded at an unprecedented rate compared to the uptake of conventional hybrids. Farmers had strong expectations on the profitability of GM crops, in particular as regards yield and/or cost savings.

Many surveys and studies have been carried out to assess reasons for adopting GM crops. They have confirmed that adoption of GM crops by farmers has been driven by profitability expectations. According to an USDA survey, the majority of farmers cited increase in yield as first reason for adoption. Savings in costs appear to be the second reason, mentioned by 20 to 40% of the respondents. This survey was conducted in 1997, only one or two years after the introduction of the first GM seeds on the US market.

In practice, the most immediate and tangible ground for satisfaction appears to be the combined effect of performance (not necessarily measured by yields) and convenience of GM crops, in particular for herbicide tolerant varieties. These crops allow for a greater flexibility in growing practices and in given cases, for reduced or more flexible labour requirements. Where labour or time is a restriction, this convenience effect has an economic impact. In the medium term, it should translate into increased labour productivity and savings in labour costs. In the long run, it might have an impact on farm restructuring, alongside with many other factors which play a role in this process.

The effective profitability of a GM crop can only be properly assessed on the basis of several years of cultivation and commercialisation. Several years have to be considered for two main reasons. First, many other factors have an impact on profitability. In particular, there are important yearly fluctuations in yields and prices. Second, effective profitability depends on developments on the supply and on the demand side. The first generation of GM crops is input-oriented. The primary effects of this new technology were expected and observed on the supply side.

There are five possible effects of technical changes in the field of plant breeding:

1. Increase in the maximum yield
2. Increase of the economically optimal yield
3. Input-switching technical change, lowering the cost but yield neutral
4. Quality-enhancing technical change
5. Risk-reducing technical change.

Bt corn falls under category 2, while Herbicide-Tolerant soybeans rather have a type-3 effect. Both types of effects imply a shift in farmers supply functions. Under given prices, farmers produce more. If the demand function remains unchanged, prices drop. Only type-4 technological change induces a structural change in the demand function, and possible increases in prices. When assessing the profitability for farmers and the economic impact of biotechnology on agri-food markets developments in supply and in demand have to be considered together. However, it appears that this has not always been the case, neither for farmers, nor for the leading biotech firms. Their approach has been supply-oriented.

Effect of Agricultural Policy

In the US as well as in the EU, GM and non-GM crops are not treated differently under the various support schemes, both are eligible. In the US, crops for which GM varieties have rapidly developed are all eligible for support under the flexibility payments, the marketing loan system, as well as for crop insurance. Soybeans became eligible for flexibility payments and under the marketing loan system in 1996, which is the year of first commercial sowings of GM varieties. Several analysts consider that existing support systems have favoured the development of soybeans sowings. In particular, the loan rate applied to soybeans makes this crop attractive compared to wheat and corn. The area under soybeans is expected to reach a record level in 2000, while prices are low.

By mid-November 1999, the United States Department of Agriculture (USDA) estimated that 90% of the 1998 soybeans crop had received a marketing loan benefit, and that the average value of this benefit was worth around 0.44 US $/bushel (14.5 €/t). Oilseed producers are also eligible for the 1999/2000 emergency packages. A specific assistance programme was set up in early February for oilseeds producers, to offset record low market prices. Under this programme, payments for soybeans could average 0.141 US $/bushel (5.3 €/t), according to calculation by private consultants. Favourable support conditions for soybeans could have played a role in the rapid uptake of GM technology for this crop.

In addition, in a low market price context, the expectation on cost savings is a further driving force for the adoption of the technology. Eligibility of GM crops under various support schemes limits the price risk of the productivity-enhancing technology. It accounts as another reason for the farmers to focus their planting decision on expected farm-level performance, on cost-efficiency of inputs. In other words, farmers also had an input-oriented approach.

Profitability of GM and non-GM Crops

Profitability is defined as the margin left over to farmers when costs have been deduced from receipts. The profitability of a GM crop is judged against corresponding conventional crops. Comparing the performance of both types of crops raises several methodological issues.

Input-effect of GM crops

Generally the cost comparison of GM crops and their conventional counterpart is limited to crop-specific costs, assuming that fixed costs are more or less the same. GM seeds are sold at a higher price than conventional ones. The price wedge is mainly attributable to the value of GM technology or to the "technological fee". Furman Selz reports about premia observed on the US market in 1998: US $ 30 per bag of seeds for GM corn and US $ 5 for GM soybeans seeds, which represents a 30% price-premium compared to non-GM seeds. They also give an indication on the average technological fee paid by seed companies to gene providers: US $ 27 (30% of GM seed price) per bag of corn seeds and US $ 4.25 (21% of GM seed price) for soybean seeds. Despite of the technological fee, GM seeds appear to be more profitable than conventional ones for seed companies.

The effect of GM crops allows for reduced or more flexible working requirements. However, the related savings in labour costs have not always been properly assessed. The valuation of family work is rarely broken down on a crop-specific basis. On the other hand, growing GM crops requires new management skills, growing practices and possible constraints. GM seeds are generally sold and sown in the context of contracts. These changes entail transaction and management costs, which are not easy to assess.

GM crops are expected to allow for cost-savings through reduced insect and weed control and/or to achieve higher yields. Under the assumption that the price of non-GM and GM crops is the same, the latter will become more profitable for farmers if the increased seed costs are offset by savings in weed and/or pest control costs and/or by higher yields.

Yields and Prices

Yield is a key factor for profitability expectations and results. In fact, available figures on crop-specific costs are often broken down on an area basis, while prices are paid on a quantity basis. Based on yields, costs and prices are brought on a common basis, often per acre/ha. In other words, the effect of possible increase in yields is taken into account on the receipt side. Comparing yields of GM and non-GM crops is not a straightforward exercise. Yields depend on a large number of factors, and the inserted trait of GM crops is only one factor amongst others. It is worth recalling that first generation genetic modifications address production conditions (pests, weeds), they do not increase the intrinsic yield capacity of the plant.

In other words, referring to Bullock's classification, they do not induce a type 1 (maximum yield) technical change. Not surprisingly yield performance of GM crops against their non-GM counterparts depends on growing conditions, in particular on the degree of infestation in insects or in weeds. Data about yields of GM crops are widely available, however, often specifications on factors which influence yields are missing, such as temperature, weed control applied etc.

The USDA has examined different factors affecting the adoption of GM crops. These include farm size, education and experience, location of the farm, use of production or marketing contracts. In the case of herbicide resistant soybeans, the USDA has concluded that "larger operations and more educated operators are more likely to use herbicide tolerant soybean seeds". Such differences between adopters and non-adopters of biotechnology have to be taken into account when comparing yields and returns obtained on both types of farms.

Another key factor on the receipt side is the market price of GM crops. In many profitability studies, prices of GM and non-GM crops are assumed to be equivalent. Most of the available studies are based on 1997 or 1998 data. In these first years of commercialisation of GM crops, their impact on commodity prices has not been manifest or is difficult to assess. Different pricing developments between GM and non-GM crops have only been observed in 1999.

There are two main crops under cultivation, respectively Herbicide Tolerant (HT) soybeans and Insect Resistant (*Bt*) corn.

Herbicide Tolerant Soybeans

Three different types of GM soybeans have been authorised in the US. Two of them are tolerant to different herbicides. Soybeans tolerant to glyphosate, Monsanto's "Round up Ready"(RR) soybeans, have been on the market since 1996 and are the most widely grown (estimated 80% share in GM soybeans). The third one is a high oleic soybean variety.

One of the reasons for the rapid adoption of GM soybeans has been the expectation of a higher yield than for non-GM soybeans. A number of US research projects have addressed this issue. Results seem to indicate the reverse: in most field trials the GM crop shows lower yields than the non-GM crop, as indicated in the table 3, in the case of Roundup Ready (RR) soybeans.

In South Minnesota, average performance of top yielding Roundup Ready soybean varieties was 3% lower than 'the top yielding conventional varieties, yet in Central Minnesota the yield drag was as much as 13% and in Southern Wisconsin 6%. While indicating lower yields in each case, these sub-regional results point to the great variability in yield performance.

Table 3. Differences in yields between conventional and GM soybeans

States	*Yield (T/ha)*		*Difference in %* *(RR-conventional)*
	Conventional	*Roundup Ready*	
Illinois	3.90	4.04	+3.5%
Iowa	4.10	3.83	–7%
Michigan	4.44	4.30	–3%
Minnesota	4.44	4.10	–8%
Nebraska	3.90	3.43	–12%
Ohio	4.04	3.90	–3%
South Dakota	3.30	2.96	–10%
Wisconsin	4.77	4.64	–3%

The USDA estimated, on the basis of the 1997 data, that the increased use of HT soybeans produced only a small global increase in yields. One of the explanations given for the lower yield of GM crops is that the GM-traits were initially not introduced in the top yielding varieties of soybean. Seed companies are now incorporating these traits in their yield-leading varieties. If this is indeed the case, then the yield drag should diminish in the coming years.

Reduced herbicide use and costs

In the 1960s herbicide use started to replace tillage and cultivation practices as a primary means of weed control.

At that time, these were mainly pre-emergence herbicides. The use of post-emergence herbicide in the production of soybean has been rising steadily since they became available in the 80s. In 1988, 44% of soybean acres were treated, by 1994 this share had risen to 72%. Quite often, they were used in combination with pre-emergence herbicides. However these classical herbicides had a number of drawbacks:

— difficult management

— risk of crop damaged

— development of herbicide resistant weeds

— some herbicides limit the possibility of crop rotation.

The emergence of GM soybeans which are tolerant to glyphosate ("Roundup") has a significant impact on the use of other herbicides. For instance, the use of imazetaphyr ("Pursuit"), one of the most widely used post-emergence herbicides has declined from 44% of soybean acres in 1995 to 17% in 1998. The main advantages of using Roundup on HT soybeans are:

— a wider window of application, both in terms of stage of growth of soybeans and effective control of larger weeds,

— the easier management of weed control programs,

— the fact that there is no carry over, thus giving growers more rotation options.

The use of this product has increased drastically. In 1990, about 10% of all soybean acreage were treated with Roundup (at that stage used only as "burndown" treatment). This figure has risen to 45% in 1998. According to the USDA, the use of other synthetic herbicides have declined by a larger amount, and the net impact of increased cultivation of HT soybeans is a decrease in overall herbicide applications. The cost of a program of Roundup on HT soybeans was 14.7 €/acre

(36.6 €/ha) in 1998, compared to 12€/acre (29.8 €/ha) for a conventional program with pre-plant treatment alone, or 22.3€/acre (55.2 €/ha) for programs using other combinations).

However, due to emergence of resistance in the future additional treatments may be needed. From 1998 to 1999, an increase from 15 to 25% in terms of average pounds of Roundup/acre was observed. Benbrook reports an increase from 24 ounce/acre to 32-48 ounce/acre in the dose of Roundup Ultra required to gain adequate control of velvetleaf and ragweed species. This would clearly have an impact on the cost of GM crops. Nevertheless, in the short term, the cost saving effect seems to be dominant. In the Duffy report, farmers who used GM crops reported spending nearly 30% less than those who grew non-GM soybeans. Reduced herbicide costs was listed by 27% of farmers as one of the reasons for planting GM crops. Furman Selz reports a 33 to 35 €/ha lower herbicide cost for HT soybeans.

Convenience effect

It is difficult to quantify the convenience effect of choosing HT crops. However, there are some clear advantages. For example:

— The ease of the glyphosate-herbicide use and the large time window for spraying, which increases flexibility.

— HT crops make the adoption of no-till or conservation tillage easier. According to Monsanto, in 1997, nearly half of the acres planted in RR soybeans are not tilled anymore. The absence or limitation of tillage implies lower use of crop-specific resources (labour, fuel etc). It is also considered to be more environmental friendly, in particular as it reduces soil erosion.

There are two counterbalancing elements in the growing of GM soybeans. On the one hand, seed prices Of GM crops are higher while yields (and thus, in a hypothesis of the same price for both variants, income) are lower, on the other hand, input costs are lower as well. However, if HT soybeans allow for savings in labour through their convenience effect, the same return for less labour means an increase in income per working hour.

Table 4. Comparison of returns for GM and conventional soybeans

Crop	*Yield (t/ha)*	*Seed Cost (€/ha)*	*Total cost (excluding land/ labour) (€/ha)*	*Return on land/ labour (€/ha)*
GMO	3.295	57	254	320
Non-GMO	3.430	42	274	322

The costs (total costs excluding land and labour) for non-GM in the Duffy study are reported to be 8% higher than for GM crops. However, these higher costs are offset by the higher yields. Similarly, in simulations of the University of Illinois, the variable costs/acre for non-GM crops were estimated to be 6 to 8% higher than for GM crops. The USDA has found no evidence of a significant change in variable profits in 1997, following the dramatic increase in GM soybeans sowings (Table 4).

Bt corn

Bt-corn, which is the leading GM corn and has been grown on a wide basis for two or three years. By their stalk tunnelling action, corn borers are significantly damaging to corn crops. During one growing season, up to three generations of corn borers can affect a given crop. To be effective, insecticide applications have to be carried out at the appropriate stage of development. Hence they require scouting, or in other words, farmers have to follow developments regarding population and to make their applications decision on this basis, in due time. For cost/effectiveness reasons, uses of insecticide sprays against corn borers vary greatly from one production region to another, or even, from one grower to another.

A soil bacterium, the *Bacillus thuringiensis* (*Bt*) produces toxins that kill the European Corn Borer. *Bt* corn includes gene material from the *Bt* bacteria, which allows own production of insecticide during the growth stage of the plant. Hence it is expected to have a yield and convenience advantage against non-*Bt* corn. A survey carried out in Iowa has shown that 80% of *Bt*-corn growers had chosen this option because of the expected yield advantage.

Several studies have found evidence on yield gains for *Bt* corn. Based on 1996-1998 data of the Agricultural Resources Management Data, the USDA has observed that adopters of *Bt* corn had obtained higher yields than non-adopters. This might however been partly explained by performance differences between these two groups of farmers. The gap between 1997 and 1998 results can be explained by the difference in weather conditions and in insect pressure. Infestation was low in 1998.

Insecticide treatments are significantly lower for *Bt* corn than for conventional corn. Insecticides were not used at all by 50% of farmers. However, it is not clear whether the absence of applications results from *Bt* technology or if it was already the case with conventional varieties. Some farmers still spray insecticide on *Bt* corn, because its performance against second or third generation infestation is more limited. In addition, insecticide may still be needed against other pests. Considering that most of the farmers do not apply insecticide for controlling European Corn Borer (ECB), Furman Selz conclude that the value of *Bt* corn is not insecticide cost savings, but rather yield protection.

Two-tiers Crop Management

To prevent resistance in ECB populations, farmers planting *Bt* crops have been advised to keep "refuges" with non-*Bt* crops next to the *Bt*-fields. In early 2000, the US Environmental Protection Agency specified requirements which have to be observed in this respect. Refuges should cover at least 20% of the area planted in *Bt* corn. Where *Bt* corn is grown near *Bt* cotton, refuges have to cover an area equivalent to 50% of the *Bt* area. This should translate into increased cultivation constraints. It has been argued that resistance to *Bt* could raise problems for organic farming, which traditionally uses sprays or granulates of *Bt* preparations within pest control programmes.

GM seeds are more expensive than conventional ones. This reflects both the technological fee charged by some biotech firms and the fact that GM and conventional seeds are sold on different markets. As explained by Hyde et al, the profitability of Insect resistant crop will depend on whether the "value of the protection" is less or more than the highest seed price. Results obtained by this research team for Indiana suggest that this is generally not the case. However, results depend on the level of infestation. Hyde and al have found that "when the probability of infestation increases from 25 to 40%, *Bt* corn value increases by about 69%". Therefore, Hyde considers that in areas where infestation is more likely or where average yields are higher, *Bt* corn should be profitable.

Several other studies show that profitability of *Bt* is higher where infestation is high (Table 5). The calculations carried out by Furman Selz are summarised in the following table.

Table 5. Farmers economics for Bt corn, various infestation scenarios

		Degree of infestation		
Yield loss if untreated	*Units*	*Light 5%*	*Medium 10%*	*Heavy 20%*
Price	€/tonne	98.4	98.4	98.4
Yield gain	t/ha	0.471	0.941	1.883
Gain in receipts	€/ha	46.3	92.7	185.3
Additional cost	€/ha	21.8	21.8	21.8
Net gain/loss	€/ha	24.5	70.9	163.5

The cost of GM seeds is also a key factor in the relative profitability of GM crops. Alexander and Goodhue have examined the relationship between seed price and profitability, as well as the likely breakdown of profitability between firms and farmers for various types of GM corn in Iowa. They found that the ranking of net revenue performance matched the ranking of seed costs. Under their simulations, *Bt* corn appears to be the type of GM corn most likely to allow profits for farmers. A possible factor of explanation might be the number of *Bt* Corn types on the market (7 transformation events have been authorised in the US). There is a competition between these types of *Bt*, which are later incorporated into various hybrids. Their margin of manoeuvre as far as prices of GM seeds and associated agro-chemical products are concerned is a key factor in the breakdown of profitability of GM crops. Farm-gate profitability of GM crops is very sensitive to input prices.

Herbicide Tolerant Canola

Canola is a type of rapeseed which has been developed and is grown in Canada. It is a registered trademark, corresponding to specified low contents in erucic acid in oil and in glucosinolates in meals. It has initially been obtained through conventional breeding, but in recent years, GM herbicide tolerant varieties have been developed. The importance of Canola has increased drastically: barely grown twenty years ago, it became the third most important crop in Canada in 1994, its value representing 29% of all grains and oilseed receipts. Canola production in Canada is mainly limited to the provinces of Alberta, Manitoba and Saskatchewan. These three provinces produce more then 98% of the Canadian Canola output (Table 6).

Table 6. Costs and returns of different Canola varieties (in €/ha)

Type of Soil		*Black soil*		*Dark brown soil*	
Type of Canola	*HT*	*Argentine*	*Polish (*)*	*HT*	*Argentine*
Gross return	342	379	307	278	259
Crop sales receipts	328	353	296	240	255
Insur. receipts	7	5	0	29	0
Misc. receipts	4	8	7	9	2
Govt. programs	2	9	4	0	3
Straw/Grazing	0	4	0	0	0
Variable costs	182	190	185	181	184
of which seeds	36	18	36	27	21
of which fertilizer	42	44	43	36	48
of which chemicals	32	51	25	31	35
Capital Costs	75	92	102	66	63
Total Prod. costs	257	281	287	248	247
GROSS MARGIN	131	163	76	84	48
Yield (Bu/acre)	27	29	24	20	21
yield (bu/Ha)	67	71	59	50	53
Avg. area (Ha)	86	65	70	102	73

Canola yields have gone up throughout the 1980s and 1990s, for example, in the province of Ontario yield has doubled between 1983 and 1996. Yield data comparing herbicide tolerant (GM) Canola to conventional Canola does not prove to be convergent. Typically, the production of Canola requires two herbicide applications: one pre-emergent and the other post-emergent, the latter controlling only for a limited spectrum of weeds. The characteristic of herbicide resistance offered by GM Canola therefore improves potential in two ways:

— removing competition for moisture and nutrients between Canola and weed.

— eliminating costs for additional machine movements over the field.

Costs and Profitability

Table 6 illustrates the difficulty of comparing profitability of these varieties, due to the number of factors which might have an effect.

— Black soil areas allow for higher yields.

— In the dark brown zone, there is no significant difference in yields and in total costs between the Argentine and the HT variety. Differences appear on the receipt side, lower sale receipts for HT crops, but higher insurance revenue.

— In the black soil area, there are significant differences in yields, costs and receipts. The "Argentine" variety achieves the highest yield, with the HT variety coming close to that level. Total costs for HT Canola are lower than for conventional varieties, and this is mainly the result of reduced capital costs.

Although variable costs of HT and conventional Canola are broadly equivalent, even from one zone to another, the breakdown is different. While costs are higher for GM seeds, those for fertiliser and herbicides are lower for HT than for "Argentine" Canola. The convenience effect of HT Canola is reflected in the lower labour and fuel costs.

References

1. Economic Impacts of Genetically Modified Crops on the Agri-Food Sector, Working Document Rev. 2 Directorate-General for Agriculture, EUROPA.
2. Goodman, DC (1987) *From Farming to Biotechnology: A Theory of Agro-industrial Development*. Oxford: Blackwell.
3. Old, RW and Primrose, SB (1989) *Principles of Gene Manipulation-4th Edition*. London: Blackwell Scientific Publications.
4. Royal Society (2000) *Transgenic Plants and World Agriculture*. Joint Report by Brazilian Academy of Sciences, Chinese Academy of Sciences, Indian National Science Academy, Mexican Academy of Sciences, National Academy of Sciences of the USA, The Royal Society (UK) and Third World Academy of Sciences. July 2000. 20 pp.
5. Seabrook, J. (1993) "Tremors in the Hothouse." The New Yorker. July 19:32-41.
6. Serageldin I (2003) IUFoST Distinguished Lecture, 12th World Congress of Food Science and Technology, Chicago, USA, 16-20 July 2003. Commentary by K. Buckle, Newsline, 55 (July-August, 2003), pp 1-3.
7. Shand, HJ (1994) *Agricultural biotechnology and the public good*. (6): p.73-86.
8. TIFS (2003) *Genetically Modified Foods for Human Health and Nutrition: The Scientific Basis for Benefit/Risk Assessment*. Trends Food Sci. Technol. 14 (5-8): 169-338 (Special Issue, 10 chapters).

7

Managing the Platform of Biotechnology

Biotechnology, an emerging knowledge-intensive field, is a set of enabling techniques for bringing about specific man-made changes in deoxyribonucleic acid (DNA), or genetic material, in plants, animals and microbial systems, leading to useful products and technologies. By itself, biotechnology cannot resolve all the fundamental problems of environment and development, so expectations need to be tempered by realism. Nevertheless, it promises to make a significant contribution in enabling the development of, for example, better health care, enhanced food security through sustainable agricultural practices, improved supplies of potable water, more efficient industrial development processes for transforming raw materials, support for sustainable methods of afforestation and reforestation, and detoxification of hazardous wastes.

Biotechnology also offers new opportunities for global partnerships, especially between the countries rich in biological resources (which include genetic resources) but lacking the expertise and investments needed to apply such resources through biotechnology and the countries that have developed the technological expertise to transform biological resources so that they serve the needs of sustainable development.

Biotechnology can assist in the conservation of those resources through, for example, *ex situ* techniques. The programme areas set out below seek to foster internationally agreed principles to be applied to ensure the sound management of biotechnology, to engender public trust and confidence, to promote the development of sustainable applications of biotechnology and to establish appropriate enabling mechanisms, especially within developing countries, through the following activities:

(a) Increasing the availability of food, feed and renewable raw materials;

(b) Improving human health;

(c) Enhancing protection of the environment;

(d) Enhancing safety and developing international mechanisms for cooperation;

(e) Establishing enabling mechanisms for the development and the environmentally sound application of biotechnology.

Increasing the Availability of Food, Feed and Renewable Raw Materials

To meet the growing consumption needs of the global population, the challenge is not only to increase food supply, but also to improve food distribution significantly while simultaneously developing more sustainable agricultural systems. Much of this increased productivity will need to take place in developing countries. It will require the successful and environmentally safe application of biotechnology in agriculture, in the environment and in human health care. Most of the investment in modern biotechnology has been in the industrialized world. Significant new investments and human resource development will be required in biotechnology, especially in the developing world.

The following objectives are proposed, keeping in mind the need to promote the use of appropriate safety measures based on programme area:

(a) To increase to the optimum possible extent the yield of major crops, livestock, and aquaculture species, by using the combined resources of modern biotechnology and conventional plant/animal/micro-organism improvement, including the more diverse use of genetic material resources, both hybrid and original.

(b) To reduce the need for volume increases of food, feed and raw materials by improving the nutritional value (composition) of the source crops, animals and micro-organisms, and to reduce post-harvest losses of plant and animal products;

(c) To increase the use of integrated pest, disease and crop management techniques to eliminate overdependence on agrochemicals, thereby encouraging environmentally sustainable agricultural practices;

(d) To evaluate the agricultural potential of marginal lands in comparison with other potential uses and to develop, where appropriate, systems allowing for sustainable productivity increases;

(e) To expand the applications of biotechnology in forestry, both for increasing yields and more efficient utilization of forest products and for improving afforestation and reforestation techniques. Efforts should be concentrated on species and products that are grown in and are of value particularly for developing countries;

(f) To increase the efficiency of nitrogen fixation and mineral absorption by the symbiosis of higher plants with micro-organisms;

(g) To improve capabilities in basic and applied sciences and in the management of complex interdisciplinary research projects.

Management-related Activities

Governments at the appropriate level, with the assistance of international and regional organizations and with the support of non-governmental organizations, the private sector and academic and scientific institutions, should improve both plant and animal breeding and micro-organisms through the use of traditional and modern biotechnologies, to enhance sustainable agricultural output to achieve food security, particularly in developing countries, with due regard to the prior identification of desired characteristics before modification, taking into account the needs of farmers, the socio-economic, cultural and environmental impacts of modifications and the need to promote sustainable social and economic development, paying particular attention to how the use of biotechnology will impact on the maintenance of environmental integrity.

More specifically, these entities should:

(a) Improve productivity, nutritional quality and shelf-life of food and animal feed products, with efforts including work on pre-and post-harvest losses;

(b) Further develop resistance to diseases and pests;

(c) Develop plant cultivars tolerant and/or resistant to stress from factors such as pests and diseases and from abiotic causes;

(d) Promote the use of underutilized crops of possible future importance for human nutrition and industrial supply of raw materials;

(e) Increase the efficiency of symbiotic processes that assist sustainable agricultural production;

(f) Facilitate the conservation and safe exchange of plant, animal and microbial germ plasm by applying risk assessment and management procedures, including improved diagnostic techniques for detection of pests and diseases by better methods of rapid propagation;

(g) Develop improved diagnostic techniques and vaccines for the prevention and spread of diseases and for rapid assessment of toxins or infectious organisms in products for human use or livestock feed;

(h) Identify more productive strains of fast-growing trees, especially for fuel wood, and develop rapid propagation methods to aid their wider dissemination and use;

(i) Evaluate the use of various biotechnology techniques to improve the yields of fish, algal and other aquatic species;

(j) Promote sustainable agricultural output by strengthening and broadening the capacity and scope of existing research centres to achieve the necessary critical mass through encouragement and monitoring of research into the development of biological products and processes of productive and environmental value that are economically and socially feasible, while taking safety considerations into account;

(k) Promote the integration of appropriate and traditional biotechnologies for the purposes of cultivating genetically modified plants, rearing healthy animals and protecting forest genetic resources;

(l) Develop processes to increase the availability of materials derived from biotechnology for use in food, feed and renewable raw materials production.

The following activities should be undertaken:

(a) Consideration of comparative assessments of the potential of the different technologies for food production, together with a system for assessing the possible effects of biotechnologies on international trade in agricultural products;

(b) Examination of the implications of the withdrawal of subsidies and the possible use of other economic instruments to reflect the environmental costs associated with the unsustainable use of agrochemicals;

(c) Maintenance and development of data banks of information on environmental and health impacts of organisms to facilitate risk assessment;

(d) Acceleration of technology acquisition, transfer and adaptation by developing countries to support national activities that promote food security.

International and Regional Cooperation and Coordination

Governments at the appropriate level, with the support of relevant international and regional organizations, should promote the following activities in conformity with international agreements or Arrangements on biological diversity, as appropriate:

(a) Cooperation on issues related to conservation of, access to and exchange of germ plasm; rights associated with intellectual property and informal innovations, including farmers' and breeders' rights; access to the benefits of biotechnology; and bio-safety;

(b) Promotion of collaborative research programmes, especially in developing countries, to support activities outlined in this programme area, with particular reference to cooperation with local and indigenous people and their communities in the conservation of biological diversity and sustainable use of biological resources, as well as the fostering of traditional methods and knowledge of such groups in connection with these activities;

(c) Acceleration of technology acquisition, transfer and adaptation by developing countries to support national activities that promote food security, through the development of systems for substantial and sustainable productivity increases that do not damage or endanger local ecosystems.

Financing and Cost Evaluation

The Conference secretariat has estimated the average total annual cost of implementing the activities of this programme to be about $5 billion, including about $50 million from the international community on grant or concessional terms. These are indicative and order-of-magnitude estimates only and have not been reviewed by Governments.

Training of competent professionals in the basic and applied sciences at all levels (including scientific personnel, technical staff and extension workers) is one of the most essential components of any programme of this kind. Creating awareness of the benefits and risks of biotechnology is essential. Given the importance of good management of research resources for the successful completion of large multidisciplinary projects, continuing programmes of formal training for scientists should include managerial training.

Training programmes should also be developed, within the context of specific projects, to meet regional or national needs for comprehensively trained personnel capable of using advanced technology to reduce the "brain drain" from developing to developed countries. Emphasis should be given to encouraging collaboration between and training of scientists, extension workers and users to produce integrated systems. Additionally, special consideration should be given to the execution of programmes for training and exchange of knowledge on traditional biotechnologies and for training on safety procedures.

Improving Human Health

The improvement of human health is one of the most important objectives of development. The deterioration of environmental quality, notably air, water and soil pollution owing to toxic chemicals, hazardous wastes, radiation and other sources, is a matter of growing concern. This degradation of the environment resulting from inadequate or inappropriate development has a direct negative effect on human health. Malnutrition, poverty, poor human settlements, lack of good-quality potable water and inadequate sanitation facilities add to the problems of communicable and non-communicable diseases. As a consequence, the health and well-being of people are exposed to increasing pressures.

The main objective of this programme area is to contribute, through the environmentally sound application of biotechnology to an overall health programme, to:

(a) Reinforce or inaugurate (as a matter of urgency) programmes to help combat major communicable diseases;

(b) Promote good general health among people of all ages;

(c) Develop and improve programmes to assist in specific treatment of and protection from major non-communicable diseases;

(d) Create enhanced capabilities for carrying out basic and applied research and for managing interdisciplinary research.

Management-related Activities

Governments at the appropriate level, with the assistance of international and regional organizations, academic and scientific institutions, and the pharmaceutical industry, should, taking into account appropriate safety and ethical considerations:

(a) Develop national and international programmes for identifying and targeting those populations of the world most in need of improvement in general health and protection from diseases;

(b) Develop criteria for evaluating the effectiveness and the benefits and risks of the proposed activities;

(c) Establish and enforce screening, systematic sampling and evaluation procedures for drugs and medical technologies, with a view to barring the use of those that are unsafe for the purposes of experimentation; ensure that drugs and technologies relating to reproductive health are safe and effective and take account of ethical considerations;

(d) Improve, systematically sample and evaluate drinking-water quality by introducing appropriate specific measures, including diagnosis of water-borne pathogens and pollutants;

(e) Develop and make widely available new and improved vaccines against major communicable diseases that are efficient and safe and offer protection with a minimum number of doses, including intensifying efforts directed at the vaccines needed to combat common diseases of children;

(f) Develop biodegradable delivery systems for vaccines that eliminate the need for present multiple-dose schedules, facilitate better coverage of the population and reduce the costs of immunization;

(g) Develop effective biological control agents against disease-transmitting vectors, such as mosquitoes and resistant variants, taking account of environmental protection considerations;

(h) Using the tools provided by modern biotechnology, develop, inter alia, improved diagnostics, new drugs and improved treatments and delivery systems;

(i) Develop the improvement and more effective utilization of medicinal plants and other related sources;

(j) Develop processes to increase the availability of materials derived from biotechnology, for use in improving human health.

The following activities should be undertaken:

a) Research to assess the comparative social, environmental and financial costs and benefits of different technologies for basic and reproductive health care within a framework of universal safety and ethical considerations;

b) Development of public education programmes directed at decision makers and the general public to encourage awareness and understanding of the relative benefits and risks of modern biotechnology, according to ethical and cultural considerations.

International and Regional Cooperation and Coordination

Governments at the appropriate levels, with the support of relevant international and regional organizations, should:

(a) Develop and strengthen appropriate safety procedures based on programme area, taking account of ethical considerations;

(b) Support the development of national programmes, particularly in developing countries, for improvements in general health, especially protection from major communicable diseases, common diseases of children and disease-transmitting factors.

To achieve the above goals, the activities need to be implemented with urgency if progress towards the control of major communicable diseases is to be achieved by the beginning of the next century. The spread of some diseases to all regions of the world calls for global measures. For more localized diseases, regional or national policies will be more appropriate. The achievement of goals calls for:

(a) Continuous international commitment;

(b) National priorities with a defined time-frame;

(c) Scientific and financial input at global and national levels.

Financing and cost evaluation

The Conference secretariat has estimated the average total annual cost (1993-2000) of implementing the activities of this programme to be about $14 billion, including about $130 million from the international community on grant or concessional terms. These are indicative and order-of-magnitude estimates only and have not been reviewed by Governments. Actual costs and financial terms, including any that are non-concessional, will depend upon, inter alia, the specific strategies and programmes Governments decide upon for implementation.

Scientific and technological means

Well-coordinated multidisciplinary efforts involving cooperation between scientists, financial institutions and industries will be required. At the global level, this may mean collaboration between research institutions in different countries, with funding at the intergovernmental level, possibly supported by similar collaboration at the national level. Research and development support will also need to be strengthened, together with the mechanisms for providing the transfer of relevant technology.

Human Resource Development

Training and technology transfer is needed at the global level, with regions and countries having access to, and participation in exchange of, information and expertise, particularly indigenous or traditional knowledge and related biotechnology. It is essential to create or enhance endogenous capabilities in developing countries to enable them to participate actively in the processes of biotechnology production. The training of personnel could be undertaken at three levels:

(a) That of scientists required for basic and product-oriented research;

(b) That of health personnel (to be trained in the safe use of new products) and of science managers required for complex intermultidisciplinary research;

(c) That of tertiary-level technical workers required for delivery in the field.

Enhancing Protection of the Environment

Environmental protection is an integral component of sustainable development. The environment is threatened in all its biotic and abiotic components: animals, plants, microbes and ecosystems comprising biological diversity; water, soil and air, which form the physical components of habitats and ecosystems; and all the interactions between the components of biodiversity and their sustaining habitats and ecosystems. With the continued increase in the use of chemicals, energy and non-renewable resources by an expanding global population, associated environmental problems will also increase. Despite increasing efforts to prevent waste accumulation and to promote recycling, the amount of environmental damage caused by overconsumption, the quantities of waste generated and the degree of unsustainable land use appear likely to continue growing.

The need for a diverse genetic pool of plant, animal and microbial germ plasm for sustainable development is well established. Biotechnology is one of many tools that can play an important role in supporting the rehabilitation of degraded ecosystems and landscapes. This may be done through the development of new techniques for reforestation and afforestation, germ plasm conservation, and cultivation of new plant varieties. Biotechnology can also contribute to the study of the effects exerted on the remaining organisms and on other organisms by organisms introduced into ecosystems.

The aim of this programme is to prevent, halt and reverse environmental degradation through the appropriate use of biotechnology in conjunction with other technologies, while supporting safety procedures as an integral component of the programme. Specific objectives include the inauguration as soon as possible of specific programmes with specific targets:

(a) To adopt production processes making optimal use of natural resources, by recycling biomass, recovering energy and minimizing waste generation;

(b) To promote the use of biotechnologies, with emphasis on bio-remediation of land and water, waste treatment, soil conservation, reforestation, afforestation and land rehabilitation;

(c) To apply biotechnologies and their products to protect environmental integrity with a view to long-term ecological security.

Governments at the appropriate level, with the support of relevant international and regional organizations, the private sector, non-governmental organizations and academic and scientific institutions, should:

(a) Develop environmentally sound alternatives and improvements for environmentally damaging production processes;

(b) Develop applications to minimize the requirement for unsustainable synthetic chemical input and to maximize the use of environmentally appropriate products, including natural products;

(c) Develop processes to reduce waste generation, treat waste before disposal and make use of biodegradable materials;

(d) Develop processes to recover energy and provide renewable energy sources, animal feed and raw materials from recycling organic waste and biomass;

e) Develop processes to remove pollutants from the environment, including accidental oil spills, where conventional techniques are not available or are expensive, inefficient or inadequate;

(f) Develop processes to increase the availability of planting materials, particularly indigenous varieties, for use in afforestation and reforestation and to improve sustainable yields from forests;

(g) Develop applications to increase the availability of stress-tolerant planting material for land rehabilitation and soil conservation;

(h) Promote the use of integrated pest management based on the judicious use of bio-control agents;

(i) Promote the appropriate use of bio-fertilizers within national fertilizer programmes;

(j) Promote the use of biotechnologies relevant to the conservation and scientific study of biological diversity and the sustainable use of biological resources;

(k) Develop easily applicable technologies for the treatment of sewage and organic waste;

(l) Develop new technologies for rapid screening of organisms for useful biological properties;

(m) Promote new biotechnologies for tapping mineral resources in an environmentally sustainable manner.

International and Regional Cooperation and Coordination

Governments at the appropriate level, with the support of relevant international and regional organizations, should:

(a) Strengthen research, training and development capabilities, particularly in developing countries, to support the activities outlined in this programme area;

(b) Develop mechanisms for scaling up and disseminating environmentally sound biotechnologies of high environmental importance, especially in the short term, even though those biotechnologies may have limited commercial potential;

(c) Enhance cooperation, including transfer of biotechnology, between participating countries for capacity-building.

Financing and cost evaluation

The Conference secretariat has estimated the average total annual cost (1993-2000) of implementing the activities of this programme to be about $1 billion, including about $10 million from the international community on grant or concessional terms. These are indicative and order-of-magnitude estimates only and have not been reviewed by Governments. Actual costs and financial terms, including any that are non-concessional, will depend upon, inter alia, the specific strategies and programmes Governments decide upon for implementation.

The activities for this programme area will increase the demand for trained personnel. Support for existing training programmes needs to be increased, for example, at the university and technical institute level, as well as the exchange of trained personnel between countries and regions. New and additional training programmes also need to be developed, for example, for technical and support personnel. There is also an urgent need to improve the level of understanding of biological principles and their policy implications among decision makers in Governments, and financial and other institutions.

Enhancing safety and developing international mechanisms for cooperation

There is a need for further development of internationally agreed principles on risk assessment and management of all aspects of biotechnology, which should build upon those developed at the national level. Only when adequate and transparent safety and border-control procedures are in place will the community at large be able to derive maximum benefit from, and be in a much better position to accept the potential benefits and risks of, biotechnology. Several fundamental principles could underlie many of these safety procedures, including primary consideration of the organism, building on the principle of familiarity, applied in a flexible framework, taking into account national requirements and recognizing that the logical progression is to start with a step-by-step and case-by-case approach, but also recognizing that experience has shown that in many instances a more comprehensive approach should be used, based on the experiences of the first period, leading, inter alia, to streamlining and categorizing; complementary consideration of risk assessment and risk management; and classification into contained use or release to the environment.

The aim of this programme area is to ensure safety in biotechnology development, application, exchange and transfer through international agreement on principles to be applied on risk assessment and management, with particular reference to health and environmental considerations, including the widest possible public participation and taking account of ethical considerations.

The proposed activities for this programme area call for close international cooperation. They should build upon planned or existing activities to accelerate the environmentally sound application of biotechnology, especially in developing countries.

Management-related Activities

Governments at the appropriate level, with the support of relevant international and regional organizations, the private sector, non-governmental organizations and academic and scientific institutions, should:

(a) Make the existing safety procedures widely available by collecting the existing information and adapting it to the specific needs of different countries and regions;

(b) Further develop, as necessary, the existing safety procedures to promote scientific development and categorization in the areas of risk assessment and risk management (information requirements; databases; procedures for assessing risks and conditions of release; establishment of safety conditions; monitoring and inspections, taking account of ongoing national, regional and international initiatives and avoiding duplication wherever possible);

(c) Compile, update and develop compatible safety procedures into a framework of internationally agreed principles as a basis for guidelines to be applied on safety in biotechnology, including consideration of the need for and feasibility of an international agreement, and promote information exchange as a basis for further development, drawing on the work already undertaken by international or other expert bodies;

(d) Undertake training programmes at the national and regional levels on the application of the proposed technical guidelines;

(e) Assist in exchanging information about the procedures required for safe handling and risk management and about the conditions of release of the products of biotechnology, and cooperate in providing immediate assistance in cases of emergencies that may arise in conjunction with the use of biotechnology products.

International and Regional Cooperation and Coordination

Governments at the appropriate level, with the support of the relevant international and regional organizations, should raise awareness of the relative benefits and risks of biotechnology.

Establishing enabling mechanisms for the development and the environmentally sound application of biotechnology

The accelerated development and application of biotechnologies, particularly in developing countries, will require a major effort to build up institutional capacities at the national and regional levels. In developing countries, enabling factors such as training capacity, know-how, research and development facilities and funds, industrial building capacity, capital (including venture capital) protection of intellectual property rights, and expertise in areas including marketing research, technology assessment, socio-economic assessment and safety assessment are frequently inadequate. Efforts will therefore need to be made to build up capacities in these and other areas and to match such efforts with appropriate levels of financial support.

There is therefore a need to strengthen the endogenous capacities of developing countries by means of new international initiatives to support research in order to speed up the development and application of both new and conventional biotechnologies to serve the needs of sustainable development at the local, national and regional levels. National mechanisms to allow for informed comment by the public with regard to biotechnology research and application should be part of the process.

There is a need for a much more cohesive and coordinated approach to harness available resources in the most effective manner. As with most new technologies, research in biotechnology and the application of its findings could have significant positive and negative socio-economic as well as cultural impacts. These impacts should be carefully identified in the earliest phases of the development of biotechnology in order to enable appropriate management of the consequences of transferring biotechnology.

The objectives are as follows:

(a) To promote the development and application of biotechnologies, with special emphasis on developing countries, by:

i. Enhancing existing efforts at the national, regional and global levels;

ii. Providing the necessary support for biotechnology, particularly research and product development, at the national, regional and international levels;

iii. Raising public awareness regarding the relative beneficial aspects of and risks related to biotechnology, to contribute to sustainable development;

iv. Helping to create a favourable climate for investments, industrial capacity-building and distribution/marketing;

v. Encouraging the exchange of scientists among all countries and discouraging the "brain drain";

vi. Recognizing and fostering the traditional methods and knowledge of indigenous peoples and their communities and ensuring the opportunity for their participation in the economic and commercial benefits arising from developments in biotechnology.

(b) To identify ways and means of enhancing current efforts, building wherever possible on existing enabling mechanisms, particularly regional, to determine the precise nature of the needs for additional initiatives, particularly in respect of developing countries, and to develop appropriate response strategies, including proposals for any new international mechanisms;

(c) To establish or adapt appropriate mechanisms for safety appraisal and risk assessment at the local, regional and international levels, as appropriate.

Management-related Activities

Governments at the appropriate level, with the support of international and regional organizations, the private sector, non-governmental organizations and academic and scientific institutions, should:

(a) Develop policies and mobilize additional resources to facilitate greater access to the new biotechnologies, particularly by and among developing countries;

(b) Implement programmes to create greater awareness of the potential and relative benefits and risks of the environmentally sound application of biotechnology among the public and key decision makers;

(c) Undertake an urgent review of existing enabling mechanisms, programmes and activities at the national, regional and global levels to identify strengths, weaknesses and gaps, and to assess the priority needs of developing countries;

(d) Undertake an urgent follow-up and critical review to identify ways and means of strengthening endogenous capacities within and among developing countries for the environmentally sound application of biotechnology, including, as a first step, ways to improve existing mechanisms, particularly at the regional level, and, as a subsequent step, the consideration of possible new international mechanisms, such as regional biotechnology centres;

(e) Develop strategic plans for overcoming targeted constraints by means of appropriate research, product development and marketing;

(f) Establish additional quality-assurance standards for biotechnology applications and products, where necessary.

The following activities should be undertaken: facilitation of access to existing information dissemination systems, especially among developing countries; improvement of such access where appropriate; and consideration of the development of a directory of information.

International and Regional Cooperation and Coordination

Governments at the appropriate level, with the assistance of international and regional organizations, should develop appropriate new initiatives to identify priority areas for research based on specific problems and facilitate access to new biotechnologies, particularly by and among developing countries, among relevant undertakings within those countries, in order to strengthen endogenous capacities and to support the building of research and institutional capacity in those countries.

The Conference secretariat has estimated the average total annual cost (1993-2000) of implementing the activities of this programme to be about $5 million from the international community on grant or concessional terms. These are indicative and order-of-magnitude estimates only and have not been reviewed by

Governments. Actual costs and financial terms, including any that are non-concessional, will depend upon, inter alia, the specific strategies and programmes Governments decide upon for implementation.

Workshops, symposia, seminars and other exchanges among the scientific community at the regional and global levels, on specific priority themes, will need to be organized, making full use of the existing scientific and technological manpower in each country for bringing about such exchanges.

Personnel development needs will need to be identified and additional training programmes developed at the national, regional and global levels, especially in developing countries. These should be supported by increased training at all levels, graduate, postgraduate and post-doctoral, as well as by the training of technicians and support staff, with particular reference to the generation of trained manpower in consultant services, design, engineering and marketing research.

Training programmes for lecturers training scientists and technologists in advanced research institutions in different countries throughout the world will also need to be developed, and systems giving appropriate rewards, incentives and recognition to scientists and technologists will need to be instituted. Conditions of service will also need to be improved at the national level in developing countries to encourage and nurture trained manpower with a view to retaining that manpower locally. Society should be informed of the social and cultural impact of the development and application of biotechnology.

Biotechnology research and development is undertaken both under highly sophisticated conditions and at the practical level in many countries. Efforts will be needed to ensure that the necessary infrastructure facilities for research, extension and technology activities are available on a decentralized basis. Global and regional collaboration for basic and applied research and development will also need to be further enhanced and every effort should be made to ensure that existing national and regional facilities are fully utilized. Such institutions already exist in some countries and it should be possible to make use of them for training purposes and joint research projects. Strengthening of universities, technical schools and local research institutions for the development of biotechnologies and extension services for their application will need to be developed, especially in developing countries.

References

1. Ausubel, FM *et al.* (1989) Current Protocols in Molecular Biology. New York: John Wiley and Sons.
2. *Kern M, Biotechnology and world food security*, Center for International Development, Harvard University.
3. Moravec, Hans *Mind Children: The Future of Robot and Human Intelligence*. Cambridge: Harvard University Press, 1988.
4. Panicali, D and Paoletti, E (1982) Construction of pox viruses as cloning vectors: Insertion of the thymidine kinase gene from herpes simplex virus into the DNA of infectious vaccinia virus. Proc. Natl. Acad. Sci. 79:4927-4931.
5. Rasmussin and Matheson eds., "A Sourcebook of Biotechnology Activities". National Association of Biology Teachers, 11250 Roger Bacon Drive #19, Reston, VA 22090, 1990.
6. Torrey, JG (1987) "The Development of Plant Biotechnology." American Scientist. 73:354-363.

8

Deciphering Bioinformatics for Biotechnology

In the last few decades, advances in molecular biology and the equipment available for research in this field have allowed the increasingly rapid sequencing of large portions of the genomes of several species. In fact, to date, several bacterial genomes, as well as those of some simple eukaryotes have been sequenced in full. The Human Genome Project, designed to sequence all 24 of the human chromosomes, is also progressing. Popular sequence databases, such as GenBank and EMBL, have been growing at exponential rates. This deluge of information has necessitated the careful storage, organization and indexing of sequence information. Information science has been applied to biology to produce the field called Bioinformatics.

The simplest tasks used in bioinformatics concern the creation and maintenance of databases of biological information. Nucleic acid sequences comprise the majority of such databases. While the storage and or ganization of millions of nucleotides is far from trivial, designing a database and developing an interface whereby researchers can both access existing information and submit new entries is only the beginning. The most pressing tasks in bioinformatics involve the analysis of sequence information. Computational biology is the name given to this process, and it involves the following:

— Finding the genes in the DNA sequences of various organisms

— Developing methods to predict the structure and/or function of newly discovered proteins and structural RNA sequences.

— Clustering protein sequences into families of related sequences and the development of protein models.

— Aligning similar proteins and generating phylogenetic trees to examine evolutionary relationships.

The process of evolution has produced DNA sequences that encode proteins with very specific functions. It is possible to predict the three-dimensional structure of a protein using algorithms that have been derived from our knowledge of physics, chemistry and most importantly, from the analysis of other proteins with similar amino acid sequences. The diagram below summarizes the process by which DNA sequences are used to model protein structure.

Major research efforts in bioinformatics include sequence alignment, gene finding, genome assembly, protein structure alignment, protein structure prediction, prediction of gene expression and protein-protein interactions, and the modeling of evolution. The terms bioinformatics and computational biology are often used interchangeably, although the latter typically focuses on algorithm development and specific computational methods. A common thread in projects in bioinformatics and computational biology is the use of mathematical tools to extract useful information from noisy data produced by high-throughput biological techniques. Representative problems in computational biology include the assembly of high-quality DNA sequences from fragmentary "shotgun" DNA sequencing, and the prediction of gene regulation with data from mRNA microarrays or mass spectrometry.

Research Areas of Bioinformatics

Sequence Analysis

Since the Phage ϕ-X174; was sequenced in 1977, the DNA sequence of more and more organisms has been decoded and stored in electronic databases. This data is analyzed to determine genes that code for proteins, as well as regulatory sequences. A comparison of genes within a species or between different species can show similarities between protein functions, or relations between species (the use of molecular systematics to construct phylogenetic trees).

With the growing amount of data, it long ago became impractical to analyze DNA sequences manually. Today, computer programs are used to search the genome of thousands of organisms, containing billions of nucleotides. These programs can compensate for mutations (exchanged, deleted or inserted bases) in the DNA sequence, in order to identify sequences that are related, but not identical.

A variant of this sequence alignment is used in the sequencing process itself. The so-called shotgun sequencing technique does not give a sequential list of nucleotides, but instead the sequences of thousands of small DNA fragments (each about 600-800 nucleotides long). The ends of these fragments overlap and, when aligned in the right way, make up the complete genome. Shotgun sequencing yields sequence data quickly, but the task of assembling the fragments can be quite complicated for larger genomes. In the case of the Human Genome Project, it took several months of CPU time (on a circa-2000 vintage DEC Alpha computer) to assemble the fragments. Shotgun sequencing is the method of choice for virtually all genomes sequenced today, and genome assembly algorithms are a critical area of bioinformatics research.

Another aspect of bioinformatics in sequence analysis is the automatic search for genes and regulatory sequences within a genome. Not all of the nucleotides within a genome are genes. Within the genome of higher organisms, large parts of the DNA do not serve any obvious purpose. This so-called junk DNA may, however, contain unrecognized functional elements. Bioinformatics helps to bridge the gap between genome and proteome projects, for example in the use of DNA sequence for protein identification.

Genome Annotation

In the context of genomics, annotation is the process of marking the genes and other biological features in a DNA sequence. The first genome annotation software system was designed in 1995 by Owen White, who was part of the team that sequenced and analyzed the first genome of a free-living organism to be decoded, the bacterium Haemophilus influenzae. Dr. White built a software system to find the genes, the transfer RNA, and other features, and to make initial assignments of function to those genes. Most current genome annotation systems work similarly, but the programs available for analysis of genomic DNA are constantly changing and improving. The Ensembl system is a genome annotation pipeline for the human genome developed by Ewan Birney at The Sanger Institute near Cambridge, England.

Computational Evolutionary Biology

Evolutionary biology is the study of the origin and descent of species, as well as their change over time. Recent developments in genome sequencing and the ubiquity of fast computers enable researchers to trace evolution of species by tracing changes in their DNA.

CEB research from the pre-genome era involved building computational models of populations and watching their behavior over time. The field of genetic algorithms might be described as the rough inverse of CEB—rather than investigating evolution through computer programs, it aims to improve computer programs through evolutionary principles.

Gene Expression Analysis

The expression of many genes can be determined by measuring mRNA levels with multiple techniques including microarrays, expressed cDNA sequence tag (EST) sequencing, serial analysis of gene expression (SAGE) tag sequencing, massively parallel signature sequencing (MPSS), or by measuring protein concentrations with high-throughput mass spectroscopy.

All of these techniques are extremely noise-prone and/or subject to bias in the biological measurement, and a major research area in computational biology involves developing statistical tools to separate signal from noise in high-throughput gene expression (HT) studies. HT studies are often used to determine the genes implicated in a disorder: one might compare microarray data from cancerous epithelial cells to data from non-cancerous cells to determine the proteins that cancer up-regulates and down-regulates.

Expression data is also used to infer gene regulation: one might compare microarray data from a wide variety of states of an organism to form hypotheses about the genes involved in each state. In a single-cell organism, one might compare stages of the cell cycle, along with various stress conditions (heat shock, starvation, etc.). One can then apply clustering algorithms to that expression data to determine which genes are co-expressed. Further analysis could take a variety of directions: one 2004 study analyzed the promoter sequences of co-expressed (clustered together) genes to find common regulatory elements and used machine learning techniques to identify the promoter elements involved in regulating each cluster.

Protein Expression Analysis

Protein microarrays and high throughput (HT) mass spectrometry (MS) can provide a snapshot of the proteins present in a biological sample. Bioinformatics is very much involved in making sense of protein microarray and HT MS data; the former involves a number of the same problems involve in examining microarrays targeted at mRNA, the latter involves the bioinformatics problem of matching MS data against protein sequence databases.

Analysis of Mutations in Cancer

Massive sequencing efforts are currently underway to identify point mutations in a variety of genes in cancer. The sheer volume of data produced requires automated systems to read sequence data, and to compare the sequencing results to the known sequence of the human genome, including known germline polymorphisms.

Oligonucleotide microarrays, including comparative genomic hybridization and single nucleotide polymorphism arrays, able to probe simultaneously up to several hundred thousand sites throughout the genome are being used to identify chromosomal gains and losses in cancer. Hidden Markov Model and change-point analysis methods are being developed to infer real copy number changes from often noisy data. Further informatics approaches are being developed to understand the implications of lesions found to be recurrent across many tumors.

Structure Prediction

Protein structure prediction is another important application of bioinformatics. The amino acid sequence of a protein, the so-called primary structure, can be easily determined from the sequence on the gene that codes for it. But, the protein can only function correctly if it is folded in a very special and individual way. The prediction of this folding just by looking at the amino acid sequence is quite difficult. Several methods for computer predictions of protein folding are currently under development.

One of the key ideas in bioinformatics research is the notion of homology. In the genomic branch of bioinformatics, homology is used to predict the function of a gene: if the sequence of gene A, whose function is known, is homologous to the sequence of gene B, whose function is unknown, one could infer that B may share A's function. In the structural branch of bioinformatics homology is used to determine which parts of the protein are important in structure formation and interaction with other proteins. In a technique called homology modelling, this information is used to predict the structure of a protein once the structure of a homologous protein is known. This currently remains the only way to predict protein structures reliably. One example of this is the similar protein homology between hemoglobin in humans and the hemoglobin in legumes (leghemoglobin). Both serve the same purpose of transporting oxygen in both organisms. Though both of these proteins have completely different amino acid sequences, their protein structures are virtually identical, which reflects their near identical purposes.s

Other techniques for predicting protein structure include protein threading and *de novo* (from scratch) physics-based modeling.

Preserving Biodiversity

Bioinformatics is often used for preserving biodiversity. The most important information collected is the species names, descriptions, distributions, status and size of populations, habitat needs, and how each organism interacts with other species. This information is compiled with computer databases, accessed with software programs to find, visualize, and analyze the information automatically, and most importantly, communicated to other people, especially over the internet. DNA sequences of endangered species can be preserved, and names and descriptions of specimens living in captivity are stored in order to allow as much access to the information needed to preserve biodiversity as possible.

An example this application is the Species 2000 project. It is an internet-based global research project which intends to provide information about every known species of plant, animal, fungus, and microbe in existence to be the foundation for studies of global biodiversity. Anyone in the world will be able to find vast information about any known species from an array of participating databases.

Modeling Biological Systems

Systems biology involves the use of computer simulations of cellular subsystems (such as the networks of metabolites and enzymes which comprise metabolism, signal transduction pathways and gene regulatory networks) to both analyze and visualize the complex connections of these cellular processes. Artificial life or virtual evolution attempts to understand evolutionary processes via the computer simulation of simple (artificial) life forms.

Software Tools for Computational Biology

The computational biology tool best-known among biologists is probably BLAST, an algorithm for searching large sequence (protein, DNA) databases. NCBI provides a popular implementation that searches their massive sequence databases.

Computer scripting languages such as Perl and Python are often used to interface with biological databases and parse output from bioinformatics programs. Communities of bioinformatics programmers have set up free/open source projects such as EMBOSS, Bioconductor, BioPerl, BioPython, BioRuby, and BioJava which develop and distribute shared programming tools and objects (as program modules) that make bioinformatics easier.

Creation of Sequence Databases

A sequence database is a large collection of DNA, protein, or other sequences stored on a computer. A database can include sequences from only one organism, as in databases including all the proteins in Saccharomyces cerevisiae, or it can include sequences from all organisms whose DNA has been sequenced. Sequence databases can be searched using a variety of methods. The most common is probably searching for a sequence similar to a certain target protein or gene whose sequence is already known to the user. The BLAST program is a method of this type.

A major problem with all the large genetic sequence databases is that records are deposited in them from a wide range of sources, from individual researchers to large genome sewquencing centers. As a result, the sequences themselves, and especially the biological annotations attached to these sequences, vary tremendously in quality. Also there is much redundancy, as multiple labs often submit numerous sequences that are identical, or nearly identical, to others in the databases.

Many annotations are based not on laboratory experiments, but on the results of sequence similarity searches for previously-annotated sequences. Of course, once a sequence has been annotated based on similarity to others, and itself deposited in the database, it can also become the basis for future annotations. This leads to the transitive annotation problem because there may be several such annotation transfers by sequence similarity between a particular database record and actual wet-lab experimental information. Therefore, one must always regard the biological annotations in major sequence databases with a considerable degree of skepticism, unless they can be verified by reference to published papers describing high-quality experimental data, or at least by reference to a human-curated sequence database.

Most biological databases consist of long strings of nucleotides (guanine, adenine, thymine, cytosine and uracil) and/or amino acids (threonine, serine, glycine, etc.). Each sequence of nucleotides or amino acids represents a particular gene or protein (or section thereof), respectively. Sequences are represented in shorthand, using single letter designations. This decreases the space necessary to store information and increases processing speed for analysis.

While most biological databases contain nucleotide and protein sequence information, there are also databases which include taxonmic information such as the structural and biochemical characteristics of organisms. The power and ease of using sequence information has however, made it the method of choice in modern analysis.

In the last three decades, contributions from the fields of biology and chemistry have facilitated an increase in the speed of sequencing genes and proteins. The advent of cloning technology allowed foreign DNA sequences to be easily introduced into bacteria. In this way, rapid mass production of particular DNA sequences, a necessary prelude to sequence determination, became possible. Oligonucleotide synthesis provided researchers with the ability to construct short fragments of DNA with sequences of their own choosing.

These oligonucleotides could then be used in probing vast libraries of DNA to extract genes containing that sequence. Alternatively, these DNA fragments could also be used in polymerase chain reactions to amplify existing DNA sequences or to modify these sequences. With these techniques in place, progress in biological research increased exponentially.

For researchers to benefit from all this information, however, two additional things were required:

1) ready access to the collected pool of sequence information and

2) a way to extract from this pool only those sequences of interest to a given researcher.

Simply collecting, by hand, all necessary sequence information of interest to a given project from published journal articles quickly became a formidable task. After collection, the organization and analysis of this data still remained. It could take weeks to months for a researcher to search sequences by hand in order to find related genes or proteins.

Computer technology has provided the obvious solution to this problem. Not only can computers be used to store and organize sequence information into databases, but they can also be used to analyze sequence data rapidly. The evolution of computing power and storage capacity has, so far, been able to outpace the increase in sequence information being created.

Theoretical scientists have derived new and sophisticated algorithms which allow sequences to be readily compared using probability theories. These comparisons become the basis for determining gene function, developing phylogenetic relationships and simulating protein models. The physical linking of a vast array of computers in the 1970's provided a few biologists with ready access to the expanding pool of sequence information. This web of connections, now known as the Internet, has evolved and expanded so that nearly everyone has access to this information and the tools necessary to analyze it.

Gene Finding

Gene finding is the area of computational biology that is concerned with algorithmically identifying stretches of sequence, usually genomic DNA, that are biologically functional. This especially includes protein-coding genes, but may also include other functional elements such as RNA genes and regulatory regions. Gene finding is one of the first and most important steps in understanding the genome of a species once it has been sequenced.

Determining that a sequence is functional should be distinguished from determining the function of the gene or its product. The latter still demands *in vivo* experimentation through gene knockout and other assays, although frontiers of bioinformatics research are making it increasingly possible to predict the function of a gene based on its sequence alone.

In its earliest days, "gene finding" was based on painstaking experimentation on living cells and organisms. Statistical analysis of the rates of homologous recombination of several different genes could determine their order on a certain chromosome, and information from many such experiments could be combined to create a genetic map specifying the rough location of known genes relative to each other. Today, with comprehensive genome sequence and powerful computational resources, gene finding has been redefined as a largely computational problem.

The collecting, organizing and indexing of sequence information into a database, a challenging task in itself, provides the scientist with a wealth of information, albeit of limited use. The power of a database comes not from the collection of information, but in its analysis. A sequence of DNA does not necessarily constitute a gene. It may constitute only a fragment of a gene or alternatively, it may contain several genes.

Luckily, in agreement with evolutionary principles, scientific research to date has shown that all genes share common elements. For many genetic elements, it has been possible to construct consensus sequences, those sequences best representing the norm for a given class of organisms (e.g, bacteria, eukaroytes). Common genetic elements include promoters, enhancers, polyadenylation signal sequences and protein binding sites. These elements have also been further characterized into further subelements.

Genetic elements share common sequences, and it is this fact that allows mathematical algorithms to be applied to the analysis of sequence data. A computer program for finding genes will contain at least the following elements (Table 1) :

Table 1. Elements of a gene-seeking computer program

Algorithms for pattern recognition	Probability formulae are used to determine if two sequences are statistically similar.
Data Tables	These tables contain information on consensus sequences for various genetic elements. More information enables a better analysis.
Taxonomic Differences	Consensus sequences vary between different taxonomic classes of organisms. Inclusion of these differences in an analysis speeds processing and minimizes error.
Analysis rules	These programming instructions define how algorithms are applied. They define the degree of similarity accepted and whether entire sequences and/or fragments thereof will be considered in the analysis. A good program design enables users to adjust these variables.

Extrinsic Approaches

In extrinsic gene finding systems, the target genome is searched for sequences that are similar to extrinsic evidence in the form of the known sequence of a messenger RNA (mRNA) or protein product. Given an mRNA sequence, it is trivial to derive a unique genomic DNA sequence from which it had to have been transcribed. Given a protein sequence, a family of possible coding DNA sequences can be derived by reverse translation of the genetic code. Once candidate DNA sequences have been determined, it is a relatively straightforward algorithmic problem to efficiently search a target genome for matches, complete or partial, and exact or inexact. BLAST is a widely used system designed for this purpose.

A high degree of similarity to a known messenger RNA or protein product is strong evidence that a region of a target genome is a protein-coding gene. However, to apply this approach systemically requires extensive sequencing of mRNA and protein products. Not only is this expensive, but in the cells of complex organisms, only a subset of all genes in the organism's genome are expressed at any given time, meaning that extrinsic evidence for many genes is not readily accessible in any single cell culture.

Thus, in order to collect extrinsic evidence for most or all of the genes in a complex organism, many different cell types must be studied, which itself presents further difficulties. For example, some human genes may be expressed only during development as an embryo or fetus, which might be difficult to study for ethical reasons.

Despite these difficulties, extensive transcript and protein sequence databases have been generated for human as well as other important model organisms in biology, such as mice and yeast. For example, the RefSeq database contains transcript and protein sequence from many different species, and the Ensembl system comprehensively maps this evidence to human and several other genomes. It is, however, likely that these databases are both incomplete and contain small but significant amounts of erroneous data.

Ab Initio Approaches

Because of the inherent expense and difficulty in obtaining extrinsic evidence for many genes, it is also necessary to resort to ab initio gene finding, in which genomic DNA sequence alone is systematically searched for certain telltale signs of protein-coding genes. These signs can be broadly categorized as either signals, specific sequences that indicate the presence of a gene nearby, or content, statistical properties of protein-coding sequence itself. *Ab initio* gene finding might be more accurately characterized as gene prediction, since extrinsic evidence is generally required to conclusively establish that a putative gene is functional.

In the genomes of prokaryotes, genes have specific and relatively well-understood promoter sequences (signals), such as the Pribnow box and transcription factor binding sites, which are easy to systematically identify. Also, the sequence coding for a protein occurs as one contiguous open reading frame (content), which is typically many hundred or thousands of base pairs long. The statistics of stop codons are such that even finding an open reading frame of this length is a fairly informative sign. (Since 3 of the 64 possible codons in the genetic code are stop codons, one would expect a stop codon to occur approximately every 20-25 codons, or 60-75 base pairs, in a random sequence.)

Furthermore, protein-coding DNA has certain periodicities and other statistical properties that are easy to detect in sequence of this length. These characteristics make prokaryotic gene finding relatively straightforward, and well-designed systems are able to achieve high levels of accuracy.

Ab initio gene finding in eukaryotes, especially complex organisms like humans, is considerably more challenging for several reasons. First, the promoter and other regulatory signals in these genomes are more complex and less well-understood than in prokaryotes, making them more difficult to reliably recognize. Two classic examples of signals identified by eukaryotic gene finders are CpG islands and binding sites for a poly(A) tail.

Second, splicing mechanisms employed by eukaryotic cells mean that a particular protein-coding sequence in the genome is divided into several parts (exons), separated by non-coding sequences (introns). (Splice sites are themselves another signal that eukaryotic gene finders are often designed to identify.) A typical protein-coding gene in human might be divided into a dozen exons, each less than two hundred base pairs in length, and some as short as twenty to thirty. It is therefore much more difficult to detect periodicities and other known content properties of protein-coding DNA in eukaryotes.

Advanced gene finders for both prokaryotic and eukaryotic genomes typically use complex probabilistic and computational linguistic models, especially hidden Markov models, in order to combine information from a variety of different signal and content measurements. The Glimmer system is a widely used and highly accurate gene finder for prokaryotes. Eukaryotic ab initio gene finders, by comparison, have achieved only limited success; a notable example is the GENSCAN program.

Comparative Genomics Approaches

As the entire genomes of many different species are sequenced, a promising direction in current research on gene finding is a comparative genomics approach. This is based on the principle that the forces of natural selection cause genes and other functional elements to undergo mutation at a slower rate than the rest of the genome, since mutations in functional elements are more likely to negatively impact the organism than mutations elsewhere. Genes can thus be identified by comparing the genomes of related species to detect this evolutionary pressure for conservation.

In 2003, comparative genomics analysis of several yeast species led to significant revisions of the yeast gene catalog, and similar approaches will likely lead to a significantly refined understanding of the human genome within the next few years.

Protein Modelling

There are a myriad of steps following the location of a gene locus to the realization of a three-dimensional model of the protein that it encodes.

Step One: Location of Transcription Start/Stop

A proper analysis to locate a genetic locus will usually have already pinpointed at least the approximate sites of the transcriptional start and stop. Such an analysis is usually sufficient in determining protein structure. It is the start and end codons for translation that must be determined with accuracy.

Step Two: Location of Translation Start/Stop

The first codon in a messenger RNA sequence is almost always AUG. While this reduces the number of candidate codons, the reading frame of the sequence must also be taken into consideration.

There are six reading frames possible for a given DNA sequence, three on each strand, that must be considered, unless further information is available. Since genes are usually transcribed away from their promoters, the definitive location of this element can reduce the number of possible frames to three. There is not a strong concensus between different species surrounding translation start codons. Therefore, location of the appropriate start codon will include a frame in which they are not apparent abrupt stop codons.

Knowledge of a protein's predicted molecular mass can assist this analysis. Incorrect reading frames usually predict relatively short peptide sequences. Therefore, it might seem deceptively simple to ascertain the correct frame. In bacteria, such is frequently the case. However, eukaryotes add a new obstacle to this process: introns!

Step Three: Detection of Intron/Exon Splice Sites

In eukaryotes, the reading frame is discontinuous at the level of the DNA because of the presence of introns. Unless one is working with a cDNA sequence in analysis, these introns must be spliced out and the exons joined to give the sequence that actually codes for the protein.

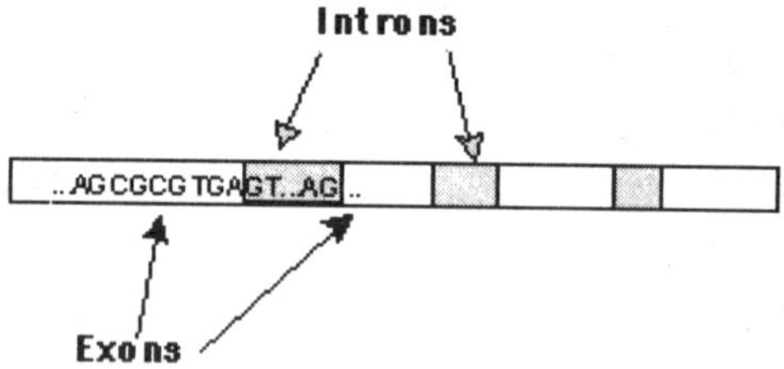

Figure 1. Intron/exon splice sites

Intron/exon splice sites can be predicted on the basis of their common features (Fig. 1). Most introns begin with the nucleotides GT and end with the nucleotides AG. There is a branch sequence near the downstream end of each intron involved in the splicing event. There is a moderate concensus around this branch site.

Step Four: Prediction of 3-D Structure

With the completed primary amino acid sequence in hand, the challenge of modelling the three-dimensional structure of the protein awaits. This process uses a wide range of data and CPU-intensive computer analysis. Most often, one is only able to obtain a rough model of the protein, and several conformations of the protein may exist that are equally probable. The best analyses will utilize data from all the following sources.

- *Pattern Comparison*: Alignment to known homologues whose conformation is more secure
- *X-ray Diffraction Data*: Most ideal when some data is available on the protein of interest. However, diffraction data from homologous proteins is also very valuable.
- *Physical Forces/Energy States*: Biophysical data and analyses of an amino acid sequence can be used to predict how it will fold in space.

All of this information is used to determine the most probable locations of the atoms of the protein in space and bond angles. Graphical programs can then use this data to depict a three-dimensional model of the protein on the two-dimensional computer screen.

Protein Structure Prediction

Protein structure prediction is one of the most significant tasks tackled in computational structural biology. It has the aim of determining the three-dimensional structure of proteins from their amino acid sequences. In more formal terms, this is the prediction of protein tertiary structure from primary structure. Given the usefulness of known protein structures in such valuable tasks as rational drug design, this is a highly active field of research.

Every two years, the performance of current methods is assessed in the CASP experiment.

The practical role of protein structure prediction is now more important than ever. Massive amounts of protein sequence data may be derived from modern large-scale DNA sequencing efforts of, for example, the Human Genome Project. The output of experimentally determined protein structures, typically by time-consuming and relatively expensive X-ray crystallography or NMR spectroscopy, is lagging far behind the output of protein sequences.

A number of factors exist that make protein structure prediction a very difficult task, including:

— The number of possible structures that proteins may possess is extremely large, as highlighted by the Levinthal paradox.

— The physical basis of protein structural stability is not fully understood.

— The primary sequence may not fully specify the tertiary structure. For example, proteins known as chaperones have the ability to induce proteins to fold in specific ways.

— Direct simulation of protein folding via methods such as molecular dynamics is not generally tractable for both practical and theoretical reasons.

Despite the above hinderances, much progress is being made by the many research groups that are interested in the task. Prediction of structures for small proteins is now a perfectly realistic goal. A wide range of approaches are routinely applied for such predictions. These approaches may be classified into two broad classes; *de novo* modelling and comparative modelling.

De novo protein modelling

De novo-or *ab initio*-protein modelling methods seek to build three-dimensional protein models "from scratch". There are many possible procedures that either attempt to mimic protein folding or apply some stochastic method to search possible solutions (i.e. global optimization of a suitable energy function). These procedures tend to require vast computational resources, and have thus only been carried out for tiny proteins.

To attempt to predict protein structure *de novo* for larger proteins, we will need better algorithms and larger computational resources like those afforded by either powerful supercomputers (such as Blue Gene) or distributed computing. Although these computational barriers are vast the potential benefits of structural genomics (by predicted or experimental methods) make *de novo* structure prediction an active research field.

Comparative protein modelling

Comparative protein modelling uses previously solved structures as starting points, or templates. This is effective because it appears that although the number of actual proteins is vast, there is a limited set of tertiary structural motifs to which most proteins belong. It has been suggested that there are only around 2000 distinct protein folds in nature, though there are many millions of different proteins.

These methods may also be split into two groups:

— Homology modelling is based on the reasonable assumption that two homologous proteins will share very similar structures. Given the amino acid sequence of a unknown structure and the solved structure of a homologous protein, each amino acid in the solved structure is mutated, computationally, into the corresponding amino acid from the unknown structure.

— Protein threading scans the amino acid sequence of an unknown structure against a database of solved structures. In each case, a scoring function is used to assess the compatibility of the sequence to the structure, thus yielding possible three-dimensional models.

Genetic Algorithm (GA)

A genetic algorithm (GA) is a search technique used in computer science to find approximate solutions to combinatorial optimization problems. Genetic algorithms are a particular class of evolutionary algorithms that use techniques inspired by evolutionary biology such as inheritance, mutation, natural selection, and recombination (or crossover).

Genetic algorithms are typically implemented as a computer simulation in which a population of abstract representations (called chromosomes) of candidate solutions (called individuals) to an optimization problem evolves toward better solutions. Traditionally, solutions are represented in binary as strings of 0s and 1s, but different encodings are also possible. The evolution starts from a population of completely random individuals and happens in generations. In each generation, the fitness of the whole population is evaluated, multiple individuals are stochastically selected from the current population (based on their fitness), modified (mutated or recombined) to form a new population, which becomes current in the next iteration of the algorithm.

Genetic algorithms or "GAs" originated from the studies of cellular automata, conducted by John Holland and his colleagues at the University of Michigan.

Research in GAs remained largely theoretical until the mid-1980s, when The First International Conference on Genetic Algorithms was held at The University of Illinois. As academic interest grew, the dramatic increase in desktop computational power allowed for practical application of the new technique. In 1989, The New York Times writer John Markoff wrote about Evolver, the first commercially available desktop genetic algorithm. Custom computer applications began to emerge in a wide variety of fields, and these algorithms are now used by a majority of Fortune 500 companies to solve difficult scheduling, data fitting, trend spotting, budgeting and virtually any other type of combinatorial optimization.

GA Operations

The problem to be solved is represented by a list of parameters which can be used to drive an evaluation procedure, called chromosomes or genomes. Chromosomes are typically represented as simple strings of data and instructions, in a manner not unlike instructions for a von Neumann machine, although a wide variety of other data structures for storing chromosomes have also been tested, with varying degrees of success in different problem domains. Initially several such parameter lists or chromosomes are generated. This may be totally random, or the programmer may seed the gene pool with "hints" to form an initial pool of possible solutions. This is called the first generation pool.

During each successive generation, each organism (or individual) is evaluated, and a value of goodness or fitness is returned by a fitness function. The pool is sorted, with those having better fitness (representing better solutions to the problem) ranked at the top. Notice that "better" in this context is relative, as initial solutions are all likely to be rather poor.

The next step is to generate a second generation pool of organisms, which is done using any or all of the genetic operators: selection, crossover (or recombination), and mutation. A pair of organisms are selected for breeding. Selection is biased towards elements of the initial generation which have better fitness, though it is usually not so biased that poorer elements have no chance to participate, in order to prevent the solution set from converging too early to a sub-optimal or local solution. There are several well-defined organism selection methods; roulette wheel selection and tournament selection are popular methods.

Following selection, the crossover (or recombination) operation is performed upon the selected chromosomes. Most genetic algorithms will have a single tweakable probability of crossover (Pc), typically between 0.6 and 1.0, which encodes the probability that two selected organisms will actually breed. A random number between 0 and 1 is generated, and if it falls below the crossover threshold, the organisms are mated; otherwise, they are propagated into the next generation unchanged.

Crossover results in two new child chromosomes, which are added to the second generation pool. The chromosomes of the parents are mixed in some way during crossover, typically by simply swapping a portion of the underlying data structure (although other, more complex merging mechanisms have proved useful for certain types of problems.) This process is repeated with different parent organisms until there are an appropriate number of candidate solutions in the second generation pool.

The next step is to mutate the newly created offspring. Typical genetic algorithms have a fixed, very small probability of mutation (Pm) of perhaps 0.01 or less. A random number between 0 and 1 is generated; if it falls within the Pm range, the new child organism's chromosome is randomly mutated in some way, typically by simply randomly altering bits in the chromosome data structure. Note however, that a very small mutation rate may lead to a de facto implementation of genetic drift (which is non-ergodic in nature) rather than to an implementation of a search algorithm that thoroughly explores the search space.

These processes ultimately result in a second generation pool of chromosomes that is different from the initial generation. Generally the average degree of fitness will have increased by this procedure for the second generation pool, since only the best organisms from the first generation are selected for breeding. The entire process is repeated for this second generation: each organism in the second generation pool is then evaluated, the fitness value for each organism is obtained, pairs are selected for breeding, a third generation pool is generated, etc. The process is repeated until an organism is produced which gives a solution that is "good enough".

A slight variant of this method of pool generation is to allow some of the better organisms from the first generation to carry over to the second, unaltered. This form of genetic algorithm is known as an elite selection strategy.

There are several general observations about the generation of solutions via a genetic algorithm:

— Unless the fitness function is handled properly, GAs may have a tendency to converge towards local optima rather than the global optimum of the problem.

— Operating on dynamic data sets is difficult, as genomes begin to converge early on towards solutions which may no longer be valid for later data. Several methods have been proposed to remedy this by increasing genetic diversity somehow and preventing early convergence, either by increasing the probability of mutation when the solution quality drops (called triggered hypermutation), or by occasionally introducing entirely new, randomly generated elements into the gene pool (called random immigrants).

— Selection is clearly an important genetic operator, but opinion is divided over the importance of crossover verses mutation. Some argue that crossover is the most important, while mutation is only necessary to ensure that potential solutions are not lost. Others argue that crossover in a largely uniform population only serves to propagate innovations originally found by mutation, and in a non-uniform population crossover is nearly always equivalent to a very large mutation.

— GAs can rapidly locate good solutions, even for difficult search spaces.

— A number of experts believe that simpler optimization algorithms can find better local optima than GAs (given the same amount of computation time). Practitioners may wish to try other algorithms in addition to GAs, especially random-restart hill climbing.

— GAs can not effectively solve problems in which there is no way to judge the fitness of an answer other than right/wrong, as there is no way to converge on the solution. These problems are often called "needle in a haystack" problems.

— As with all current machine learning problems it is worth tuning the parameters such as mutation probability and recombination probability to find reasonable setting for the problem class you are working on. There are theoretical but not yet practical upper and lower bounds for these parameters that can help guide selection.

Variants

The simplest algorithm represents each chromosome as a bit string. Typically, numeric parameters can be represented by integers, though it is possible to use floating point representations. The basic algorithm performs crossover and mutation at the bit level. Other variants treat the chromosome as a list of numbers which are indexes into an instruction table, nodes in a linked list, hashes, objects, or any other imaginable data structure. Crossover and mutation are performed so as to respect data element boundaries. For most data types, specific variation operators can be designed. Different chromosomal data types seem to work better or worse for different specific problem domains.

There have also been attempts to introduce other evolutionary operations such as movement of genes, in the manner of transposons. These movements change the schema of the chromosome making effects of linkage visible. There are also parallel implementations of genetic algorithms that use computers as 'islands' and implement migrations of populations from one computer to another over a network.

Some other variants introduce a variable fitness function. In classical genetic algorithms, the fitness function is unchanged over time. In simulated annealing the fitness function is changed over time and in artificial life, the fitness of any individual is affected by all individuals in the population with which it interacts.

Efficiency

Genetic algorithms are known to produce good results for some problem instances. Their major disadvantage is that they are relatively slow, being very computationally intensive compared to other heuristics, such as random optimization.

Recent speed improvements have focused on speciation, where crossover can only occur if individuals are closely-enough related. Genetic algorithms are extremely easy to adapt to parallel computing and clustering environments. One method simply treats each node as a parallel population. Organisms are then migrated from one pool to another according to various propagation techniques.

Another method, the Farmer/worker architecture, designates one node the farmer, responsible for organism selection and fitness assignment, and the other nodes as workers, responsible for recombination, mutation, and function evaluation.

Problem Domains

Problems which appear to be particularly appropriate for solution by genetic algorithms include timetabling and scheduling problems, and many scheduling software packages are based on GAs. GAs have also been applied to engineering. Genetic algorithms are often applied as an approach to solve global optimization problems. Genetic algorithms have been successfully applied to the study of neurological evolution.

As a general rule of thumb genetic algorithms might be useful in problem domains that have a complex fitness landscape as recombination is designed to move the population away from local minima that a traditional hill climbing algorithm might get stuck in.

Applications of GA

Spme major applications of GA are given below:

— Automated design, including research on composite material design and multi-objective design of automotive components for crashworthiness, weight savings, and other characteristics.

— Automated design of mechatronic systems using bond graphs and genetic programming (NSF).

- Calculation of Bound States and Local Density Approximations.
- Configuration applications, particularly physics applications of optimal molecule configurations for particular systems like C60 (buckyballs).
- Container loading optimization.
- Distributed computer network topologies.
- Electronic circuit design, known as Evolvable hardware.
- File allocation for a distributed system.
- Parallelization of GAs/GPs including use of hierarchical decomposition of problem domains and design spaces nesting of irregular shapes using feature matching and GAs.
- Game Theory Equilibrium Resolution.
- Learning Robot behavior using Genetic Algorithms.
- Learning fuzzy rule base using genetic algorithms.
- Mobile communications infrastructure optimization.
- Molecular Structure Optimization.
- Multiple population topologies and interchange methodologies.
- Protein folding and protein/ligand docking.
- Plant floor layout.
- Scheduling applications, including job-shop scheduling. The objective being to schedule jobs in a sequence dependent or non-sequence dependent setup environment for a minimal total tardiness.
- Solving the machine-component grouping problem required for cellular manufacturing systems.
- Tactical asset allocation and international equity strategies.

— Timetabling problems, such as designing a non-conflicting class timetable for a large university.

— Training artificial neural networks when pre-classified training examples are not readily obtainable.

— Traveling Salesman Problem.

Genetic Programming (GP)

Genetic programming (GP) is a technique popularized by John Koza, in which computer programs, rather than function parameters, are optimized. Genetic programming often uses tree-based internal data structures to represent the computer programs for adaptation instead of the list, or array, structures typical of genetic algorithms. GP algorithms typically require running time that is orders of magnitude greater than that for genetic algorithms, but they may be suitable for problems that are intractable with genetic algorithms.

Interactive genetic algorithms are genetic algorithms that use human evaluation. They are usually applied to domains where it is hard to design a computational fitness function, for example, evolving images, music, artistic designs and forms to fit users' aesthetic preference.

Simulated Annealing (SA) is a related global optimization technique which traverses the search space by testing random mutations on an individual solution. A mutation that increases fitness is always accepted. A mutation which lowers fitness is accepted probabilistically based on the difference in fitness and a decreasing temperature parameter. In SA parlance, one speaks of seeking the lowest energy instead of the maximum fitness. SA can also be used within a standard GA algorithm, simply by starting with a relatively high rate of mutation, which decreases over time along a given schedule.

References

1. *Beer MA, Tavazoie S (2004) Predicting gene expression from sequence*. In *Cell*. 2004 Apr 16;117(2):185-98.
2. Claverie JM and Notredame C (2003) *Bioinformatics for Dummies*. Wiley.
3. Durbin R, Eddy S, Krogh A and Mitchison G (1998) *Biological sequence analysis*. Cambridge University Press.
4. European Bioinformatics Institute databases (*http://www.ebi.ac.uk/Databases/*)
5. Goldberg DE (1989) *Genetic Algorithms in Search, Optimization and Machine Learning*, Kluwer Academic Publishers, Boston, MA.
6. Goldberg DE (2002) *The Design of Innovation: Lessons from and for Competent Genetic Algorithms*, Addison-Wesley, Reading, MA.
7. Harvey I (1992), *Species Adaptation Genetic Algorithms: A basis for a continuing SAGA*, in 'Toward a Practice of Autonomous Systems: Proceedings of the First European Conference on Artificial Life', F.J. Varela and P. Bourgine (eds.), MIT Press/ Bradford Books, Cambridge, MA, pp. 346-354.
8. Knowledge Center: Bioinformatics, Biomedical Informatics Ltd., March 2005.
9. Kohane, et al. (2002) *Microarrays for an Integrative Genomics*. The MIT Press.
10. Koza J (1992) *Genetic Programming: On the Programming of Computers by Means of Natural Selection*. MIT Press.
11. Michael WS (1995) *Introduction to Computational Biology: Sequences, Maps and Genomes*. CRC Press.
12. Michalewicz Z (1999) *Genetic Algorithms + Data Structures = Evolution Programs*, Springer-Verlag.
13. Mitchell M (1996) *An Introduction to Genetic Algorithms*, MIT Press, Cambridge, MA.
14. Mount DW (2002) *Bioinformatics: Sequence and Genome Analysis* Spring Harbor Press, May 2002.

15. NCBI completely sequenced genomes (*http://www.ncbi.nlm.nih.gov/entrez/query.fcgi?db=Genome*)
16. Stanford *Saccharomyces* Genome Database (*http://www.yeastgenome.org/*)
17. Vose MD (1999), *The Simple Genetic Algorithm: Foundations and Theory*, MIT Press, Cambridge, MA.

9

Biotechnology and its Noble Promises

Our way of life is likely to be more fundamentally transformed in the next several decades than in the previous one thousand years. By the year 2025, we and our children may be living in a world utterly different from anything human beings have ever experienced in the past.

In little more than a generation, our definition of life and the meaning of existence is likely to be radically altered. Long-held assumptions about nature, including our own human nature, are likely to be rethought. Many age-old practices regarding sexuality, reproduction, birth, and parenthood could be partially abandoned. Ideas about equality and democracy are also likely to be redefined, as well as our vision of what is meant by terms such as "free will" and "progress." Our very sense of self and society will likely change, as it did when the early Renaissance spirit swept over medieval Europe more than seven hundred years ago.

There are many convergent forces coming together to create this powerful new social current. At the epicenter is a technology revolution unmatched in all of history in its power to remake ourselves, our institutions, and our world. Scientists are beginning to reorganize life at the genetic level. The new tools of biology are opening up opportunities for refashioning life on Earth while foreclosing options that have existed over the millennia of evolutionary history. Before our eyes lies an uncharted new landscape whose contours are being shaped in thousands of biotechnology laboratories in universities, government agencies, and corporations around the world. If the claims already being made for the new science are only partially realized, the consequences for society and future generations are likely to be enormous. Here are just a few examples of what could happen within the next twenty-five years.

A handful of global corporations, research institutions, and governments could hold patents on virtually all 100,000 genes that make up the blueprints of the human race, as well as the cells, organs, and tissues that comprise the human body. They may also own similar patents on tens of thousands of micro-organisms, plants, and animals, allowing them unprecedented power to dictate the terms by which we and future generations will live our lives.

Global agriculture could find itself in the midst of a great transition in world history, with an increasing volume of food and fiber being grown indoors in tissue culture in giant bacteria baths, at a fraction of the price of growing staples on the land. The shift to indoor agriculture could presage the eventual elimination of the agricultural era that stretched from the neolithic revolution some ten thousand years ago, to the green revolution of the latter half of the twentieth century. While indoor agriculture could mean cheaper prices and a more abundant supply of food, millions of farmers in both the developing and developed world could be uprooted from the land, sparking one of the great social upheavals in world history.

Tens of thousands of novel transgenic bacteria, viruses, plants and animals could be released into the Earth's ecosystems for commercial tasks ranging from "bio-remediation" to the production of alternative fuels. Some of those releases, however, could wreak havoc with the planet's biosphere, spreading destabilizing and even deadly genetic pollution across the world. Military uses of the new technology might have equally devastating effects on the Earth and its inhabitants. Genetically engineered biological warfare agents could pose as serious a threat to global security in the coming century as nuclear weapons do now.

Animal and human cloning could be commonplace, with "replication" partially replacing "reproduction" for the first time in history. Genetically customized and mass-produced animal clones could be used as chemical factories to secrete in their blood and milk, large volumes of inexpensive chemicals and drugs for human use. We could also see the creation of a range of new chimeric animals on Earth, including human/animal hybrids. A chimp/hume, half chimpanzee and half human, for example, could become a reality. The human/animal hybrids could be widely used as experimental subjects in medical research and as organ "donors" for xenotransplantation. The artificial creation and propagation of cloned, chimeric, and transgenic animals could mean the end of the wild and the substitution of a bioindustrial world.

Some parents might choose to have their children conceived in test tubes and gestated in artificial wombs outside the human body to avoid the unpleasantries of pregnancy and to ensure a safe, transparent environment through which to monitor their unborn child's development. Genetic changes could be made in human fetuses in the womb to correct deadly diseases and disorders and to enhance mood, behavior, intelligence, and physical traits. Parents might be able to design some of the characteristics of their own children, fundamentally altering the very notion of parenthood. "Customized" babies could pave the way for the rise of a eugenic civilization in the twenty-first century.

Millions of people could obtain a detailed genetic readout of themselves, allowing them to gaze into their own biological futures. The genetic information would give people the power to predict and plan their lives in ways never before possible. That same "genetic information," however, could be used by schools, employers, insurance companies, and governments to determine educational tracks, employment prospects, insurance premiums, and security clearances, giving rise to a new and virulent form of discrimination based on one's genetic profile. Our notions of sociality and equity could be transformed. Meritocracy could give way to genetocracy, with individuals, ethnic groups, and races increasingly categorized and stereotyped by genotype, making way for the emergence of an informal biological caste system in countries around the world.

The Biotech Century could bring some or even most of these changes and many more into our daily lives, deeply affecting our individual and collective consciousness, the future of our civilization, and the biosphere itself. The benefits and perils of what some are calling "the ultimate technology frontier" are both exciting to behold and chilling to contemplate. Still, despite both the formidable potential and ominous nature of this extraordinary technology revolution, until now far more public attention has been focused on the other great technology revolution of the twenty-first century—computers and telecommunications. That's about to change. After more than forty years of running on parallel tracks, the information and life sciences are slowly beginning to fuse into a single technological and economic force.

The computer is increasingly being used to decipher, manage, and organize the vast genetic information that is the raw resource of the emerging biotech economy. Scientists working in the new field of "bioinformatics" are beginning to download the genetic information of millions of years of evolution, creating a powerful new genre of "biological data banks." The rich genetic information in these biological data banks is being used by researchers to remake the natural world.

The marriage of computers and genes forever asters our reality at the deepest levels of human experience. To begin to comprehend the enormity of the shift taking place in human civilization, it's important to step back and gain a better understanding of the historic nature of the many changes that are occurring around us as we turn the corner into a new century. Those changes represent a turning point for civilization. We are in the throes of one of the great transformations in world history. Before us lies the passing of one great economic era and the birth pains of another.

Millions of human beings around the world were uprooted from their rural homes and ancestral lands and forced to make their way into sprawling new urban enclaves where they sought a new kind of work in dimly lit factories and bustling offices far removed from the changing seasons and the age-old customs and rituals of an agricultural existence. Rails were laid across continents, followed in quick succession by telegraph and telephone wires and poles and miles of paved roads, changing our notions of time and distance. The Wright brothers took wings to flight and a few years later Henry Ford began supplying every American family with a standardized gasoline-powered automobile. Time zones and posted speeds heralded a new quickened pace of life, and in schools, businesses, and homes, the talk was of efficiency, the new mantra of a streamlined, hard-hitting, "future oriented" century.

Everywhere, cement foundations were laid and scaffolding went up, making way for a new vertical world of gleaming secular cathedrals made of iron, steel, aluminum, and glass. After centuries of working, living, and socializing side by side, we began a radical new experiment, living in relative isolation, one on top of another, always in search of that elusive prize which, for want of a better name, we loosely carted self-fulfillment.

It was an age of great abundance for some and growing destitution for others—giant department stores, and later, vast shopping malls were filled with exotica and trivia, staples, craft, and fine works of art. Fashion replaced necessities for millions of people. Scientists and engineers became our nwew authorities on almost everything that mattered, their views and ideas sought out, revered, elevated, and enshrined. It was the century of physics and chemistry. We peered into the micro world of atoms and electrons and rewrote the book of nature with the discoveries of quantum mechanics and relativity theory. Scientists split the atom, harnessed a new form of energy more powerful than anything human beings had ever experienced, and created the atomic bomb.

Physicists and engineers took man to the moon and back while chemists busied themselves with the creation of new kinds of more versatile synthetic materials. Plastics, a curiosity at the beginning of the century, became ubiquitous by the mid decades, seemingly as essential to our way of life as the very air we breathe. Petrochemical fertilizers and synthetic pesticides reshaped the agricultural landscape, coming just in time—claim many advocates—to help feed a growing human population that was doubling every two generations.

We built vast sewer systems underground, aerated and purified our water, improved our nutrition and increased our life expectancy by more than twenty years.

Engineers invented the X-ray machine and later the MRI. Medical researchers gave us vaccinations, anesthetics, antibiotics, and other wonder drugs.

At long last, we are nearing the end of this unique period in world history--the industrial era spread across five centuries and six continents and fundamentally changed the way human beings lived, worked and viewed themselves and their world. It was, above all else, an age propelled by cheap and abundant extractive energy. Once regarded as nearly inexhaustible, the fuel reserves of the carboniferous era are steadily diminishing, making it more costly to extract and more expensive to consume. Watching the Persian Gulf War on television night after night as hundreds of oil derricks across the Kuwaiti desert poured fire into the sky, consuming millions of barrels of precious oil for weeks and then months at a time, was a powerful reminder of just how dependent the modern world has become on these precious fuels.

Global warming is changing the very biochemistry of the planet, threatening temperature and climate changes of incalculable proportions in the coming century. Even a three-and-one-half degree Fahrenheit change in temperature brought on by global-warming gases—regarded by most scientists as a conservative forecast of what might be in store—would represent the most significant change in the Earth's climate in thousands of years. A climate change of this magnitude will likely lead to the melting of the polar ice caps, a worldwide rise in sea water level, the submerging of some island nations, the erosion of coastlines, and radical fluctuations in weather including more severe droughts, hurricanes, and tornadoes. Whole ecosystems are likely to fall victim to the radical climate shift. Agricultural regions are likely to shift far to the north, creating new opportunities for some and loss of livelihood for others.

The steady decline in fossil fuel reserves and the increasing global pollution brought on, at least in part, by the use of these same fuels is leading civilization to search for new, alternative approaches to harnessing the energy of nature in the coming century. The hard reality is that we are nearing the end of the age of fossil fuels and, with it, the end of the industrial age that has been molded from it. While the Industrial Age is not going to collapse in a fortnight or disappear in a generation or even a lifetime, its claim to the future has passed. That is not to say that the Industrial Age will not remain with us. It will, just as other great economic epochs still do. One can still travel the backwaters of the planet and stumble upon faltering pockets of neolithic and even paleolithic life.

The industrial epoch marks the final stage of the age of fire. After thousands of years of putting fire to ore, the age of pyrotechnology is slowly burning out. Fire conditioned humankind's entire existence. In the Protagoras, Plato recounts how human beings came to possess fire and the pyrotechnological arts. According to the myth, as the gods began the process of fashioning living creatures out of earth and fire, Epimetheus and Prometheus were called upon to provide them with their proper qualities. By the time they came to human beings, Prometheus noticed that Epimetheus had already distributed all the qualities at their disposal to the rest of the plants and animals. Not wanting to leave human beings totally bereft, Prometheus stole the mechanical arts and fire from the gods and gave them to men and women. With these acquisitions, humanity acquired knowledge that originally belonged only to the gods.

The age of pyrotechnology began in earnest around 3000 B.C. in the Mediterranean and Near East when people shifted from the exclusive use of muscle power to shape inanimate nature to the use of fire. Pounding, squeezing, breaking, mashing, and grinding began to give way to fusing, melting, soldering, forging, and burning. By refiring the cold remains of what was once a fireball itself, human beings began the process of recycling the crust of the planet into a new home for themselves.

Now that humanity has fashioned this second home, it finds itself in short supply of the fossil fuels necessary to keep the economic furnaces afire while increasingly vulnerable to the effects of accumulating global-warming gases that threaten to radically change the Earth's climate. The industrial way of life has also become increasingly inhospitable to the rest of the Earth's creatures, who are largely unable to adjust to this alien manmade environment. Overpopulation, logging, grazing, and land development are resulting in massive deforestation and spreading desertification. The process is leading to the extinction of many of the remaining species of life, threatening a wholesale diminution of the Earth's biological diversity upon which we rely for sources of food, fiber, and pharmaceuticals. To put the magnitude of the problem in perspective, it is estimated that during the dinosaur age, species became extinct at a rate of about one per thousand years. By the early stages of the Industrial Age, species were dying out on the average of one per decade.

Today, we are losing three species to extinction every hour. Humankind, then, faces three crises simultaneously—a dwindling of the Earth's nonrenewable energy reserves, a dangerous buildup of global-warming gases, and a steady decline in biological diversity. It is at this critical juncture that a revolutionary approach to organizing the planet is being advanced, an approach so far-reaching in scope that it will fundamentally alter humanity's relationship to the globe.

A New Operational Matrix

Great economic changes in history occur when a number of technological and social forces come together to create a new "operating matrix." There are seven strands that make up the operational matrix of the Biotech Century. Together, they create a framework for a new economic era.

First, the ability to isolate, identify, and recombine genes is making the gene pool available, for the first time, as the primary raw resource for future economic activity. Recombinant DNA techniques and other biotechnologies allow scientists and biotech companies to locate, manipulate, and exploit genetic resources for specific economic ends.

Second, the awarding of patents on genes, cell lines, genetically engineered tissue, organs, and organisms, as well as the processes used to alter them, is giving the marketplace the commercial incentive to exploit the new resources.

Third, the globalization of commerce and trade make possible the wholesale reseeding of the Earth's biosphere with a laboratory-conceived second Genesis, an artificially produced bioindustrial nature designed to replace nature's own evolutionary scheme. A global life-science industry is already beginning to wield unprecedented power over the vast biological resources of the planet. Life-science fields ranging from agriculture to medicine are being consolidated under the umbrella of giant "life" companies in the emerging biotech marketplace.

Fourth, the mapping of the approximately 100,000 genes that comprise the human genome, new breakthroughs in genetic screening, including DNA chips, somatic gene therapy, and the imminent prospect of genetic engineering of human eggs, sperm, and embryonic cells, is paving the way for the wholesale alteration of the human species and the birth of a commercially driven eugenics civilization.

Fifth, a spate of new scientific studies on the genetic basis of human behavior and the new sociobiology that favors nature over nurture are providing a cultural context for the widespread acceptance of the new biotechnologies.

Sixth, the computer is providing the communication and organizational medium to manage the genetic information that makes up the biotech economy. All over the world, researchers are using computers to decipher, download, catalogue, and organize genetic information, creating a new store of genetic capital for use in the bioindustrial age. Computational technologies and genetic technologies are fusing together into a powerful new technological reality.

Seventh, a new cosmological narrative about evolution is beginning to challenge the neo-Darwinian citadel with a view of nature that is compatible with the operating assumptions of the new technologies and the new global economy. The new ideas about nature provide the legitimizing framework for the Biotech Century by suggesting that the new way we are reorganizing our economy and society are amplifications of nature's own principles and practices and, therefore, justifiable.

The Biotech Century brings with it a new resource base, a new set of transforming technologies, new forms of commercial protection to spur commerce, a global trading market to reseed the Earth with an artificial second Genesis, an emerging eugenics science, a new supporting sociology, a new communication tool to organize and manage economic activity at the genetic level, and a new cosmological narrative to accompany the journey. Together, genes, biotechnologies, life patents, the global life-science industry, human-gene screening and surgery, the new cultural currents, computers, and the revised theories of evolution are beginning to remake our world.

The new operational matrix began to take shape in the 1950s, when biologists discovered ways of locating and identifying chromosomes and genes. In the mid 1950s, cytologists—biologists who study the workings of cells—began experimenting with ways of separating chromosomes from the rest of a cell's makeup and organizing them so they could be analyzed under a microscope. The process is called karyotyping.

In 1968, Dr. Torbjorn O. Caspersson and Dr. Lore Zech, both cytochemists at the Karolinska Institute in Sweden, invented a process for identifying chromosomes, opening the door to the mapping of genes. The researchers realized that genes have different ratios of each of the four base nucleotides G, A, T, and C. They then found a chemical, acridine quinacrine mustard, that has an affinity for the G base, and stained the chromosome with it. When placed under ultraviolet light, the chromosome "glowed in a pattern of bright and dim spots reflecting high and low concentrations of base G." Using the new banding pattern technique, Caspersson was able to identify individual human chromosomes. Other stains were subsequently developed and by the mid-1970s, researchers were studying alterations in banding patterns of chromosomes and connecting them to specific genetic traits and genetic disorders.

The new methods for isolating, identifying, and storing genes are being accompanied by a host of new techniques to manipulate and transform genes. The most formidable of the new tools is recombinant DNA. In 1973, biologists Stanley Cohen of Stanford University and Herbert Boyer of the University of California performed a feat in the world of living matter that some biotech analysts believe rivals the importance of harnessing fire. The two researchers reported taking two unrelated organisms that could not mate in nature, isolating a piece of DNA from each, and then recombining the two pieces of genetic material. A product of nearly thirty years of investigation, climaxed by a series of rapid discoveries in the late 1960s and 1970s, recombinant DNA is a kind of biological sewing machine that can be used to stitch together the genetic fabric of unrelated organisms.

The recombinant DNA process is the most dramatic technological tool to date in the growing biotechnological arsenal. The new techniques for identifying and manipulating genes are the first strand in the new operational matrix of the Biotech Century. After thousands of years of fusing, melting, soldering, forging, and burning inanimate matter to create useful things, we are now splicing, recombining, inserting, and stitching living material into economic utilities. Lord Ritchie-Calder, the British science writer, cast the biological revolution in the proper historical perspective when he observed that "just as we have manipulated plastics and metals, we are now manufacturing living materials." We are moving from the age of pyrotechnology to the age of biotechnology. The speed of the discoveries is truly phenomenal. It is estimated that biological knowledge is currently doubling every five years, and in the field of genetics, the quantity of information is doubling every twenty-four months. The commercial possibilities, say the scientists, are limited only by the span of the human imagination and the whims and caprices of the marketplace.

Efficiency and speed lie at the heart of the new genetic engineering revolution. Nature's production and recycling schedules are deemed inadequate to ensure an improved standard of living for a burgeoning human population. To compensate for nature's slower pace, new ways must be found to engineer the genetic blueprints of microbes, plants, and animals in order to accelerate their transformation into useful economic products. According to a study by the United States government's now-defunct Office of Technology Assessment, bioengineering "can play a major role in improving the speed, efficiency, and productivity of biological systems." While the motivation behind genetic engineering is age-old, the technology itself represents something qualitatively new.

We have been domesticating, breeding, and hybridizing animals and plants for more than ten millennia. But in the long history of such practices we have been restrained in what we could accomplish because of the natural constraints imposed by species borders. Although nature has, on occasion, allowed us to cross species boundaries, the incursions have always been very narrowly proscribed. Animal hybrids (mules, for example) are usually sterile, and plant hybrids do not breed true. As famed horticulturist Luther Burbank and a long line of his predecessors have understood, there are built-in limits as to how much can be manipulated when working at the organism or species level.

Genetic engineering bypasses species restraints altogether. With this new technology, manipulation occurs not at the species level but at the genetic level. The working unit is no longer the organism, but rather the gene. The implications are enormous and far-reaching. To begin with, the entire notion of a species as a separate recognizable entity with a unique nature becomes an anachronism once we begin recombining genetic traits across natural mating boundaries. Three examples illustrate the dramatic change that genetic engineering makes in our relationship to nature.

Early in 1984, a comparable feat was accomplished in England. Scientists fused together embryo cells from a goat and from a sheep, and placed the fused embryo into a surrogate animal who gave birth to a sheep-goat chimera, the first such example of the "blending" of two completely unrelated animal species in human history. In 1986, scientists took the gene whose product emits light in a firefly and inserted it into the genetic code of a tobacco plant. The tobacco leaves glow. These results could never have been achieved even with the most sophisticated conventional breeding techniques. In the biotech labs, however, the recombinant possibilities are near limitless.

The new genetic technologies allow us to combine genetic material across natural boundaries, reducing all of life to manipulatable chemical materials. This radical new form of biological manipulation changes both our concept of nature and our relationship to it. We begin to view life from the perspective of a chemist. The organism and the species no longer commands our attention or respect. Our interest now focuses increasingly on the thousands of chemical strands of genetic information that comprise the blueprints for living things.

With the newfound ability to identify, store, and manipulate the very chemical blueprints of living organisms, we assume a new role in the natural scheme of things. For the first time in history we become the engineers of life itself. We begin to reprogram the genetic codes of living things to suit our own cultural and economic needs and desires. We take on the task of creating a second Genesis, this time a synthetic one geared to the requisites of efficiency and productivity.

Today, hundreds of new bioengineering firms are setting the pace for the biotechnical revolution. In nearly every life science field, development guidelines are being laid out, long-range retooling of equipment is being hurried along, new personnel are being hired, all in a mad rush to introduce the new genetic commerce into the economy, readying civilization to taste the first fruits of the biotechnological age. There are already 1,300 biotech companies in the United States alone, with a total of nearly $13 billion in annual revenue, and more than 100,000 employees. All of this development has occurred in only the first decade of a technological and economic revolution that will likely span several centuries. The new biotechnologies are already reshaping virtually every field.

In the mining industry, researchers are developing new microorganisms that can replace the miner and his machine in the extraction of ores. As early as the 1980s, tests were being conducted with organisms that consume metals like cobalt, iron, nickel, and manganese. One company reported that it had successfully blown a certain bacterium "into low-grade copper ores where [it] produced an enzyme that eats away salts in the ore, leaving behind an almost pure form of copper." For low-grade ores that are difficult to extract with conventional mining techniques, microorganisms will provide a more economical approach to extraction and processing. For example, scientists are now using microbial agents to degrade the minerals in which gold is trapped prior to chemical extraction, to increase the recovery rate of the gold.

In the future, the mining industry is expected to turn increasingly to bioleaching with microorganisms as a more economical way to utilize low-grade ores and mineral spoils that might ordinarily be discarded. Research is even going on to design microorganisms that can consume methane gas in the mines, eliminating one of the major sources of mine explosions. Energy companies are beginning to experiment with renewable resources as a substitute for coal, oil, and natural gas. Scientists hope to improve on existing crops, like sugar cane, which is already producing fuel for automobiles.

Ethanol, derived from sugar and grain crops, is expected to provide more than 25 percent of U.S. motor vehicle fuel by the mid years of the coming century. Researchers are working on even more sophisticated approaches to biofuels, in the hope of replacing fossil fuels altogether. Scientists recently developed a strain of *E. coli* bacteria that can consume agricultural residues, yard trimmings, municipal solid waste and paper sludge, converting it to ethanol. Scientists in the chemical industry are talking about replacing petroleum, which for years has been the primary raw material for the production of plastics, with renewable resources produced by microorganisms and plants.

A British firm, ICI, has developed strains of bacteria capable of producing plastics with a range of properties, including variant degrees of elasticity. The plastic is one hundred percent biodegradable and can be used in much the same way as petrochemical-based plastic resins. Researchers are experimenting with even more exotic ways of creating new fibers and packaging materials. The United States Army is inserting genes into bacteria that are similar to the genes used by orb-weaving spiders to make silk. The spiders' silk is among the strongest fibers known to exist.

Biotechnology is also being looked to as a key tool in environmental cleanup. Bioremediation is the use of living organisms—primarily microorganisms—to remove or render harmless dangerous pollutants and hazardous waste. A new generation of genetically engineered organisms is being developed to convert toxic materials into benign substances. Researchers are using genetically engineered fungi, bacteria, and algae as "biosorption" systems to capture polluting metals and radionuclides including mercury, copper, cadmium, uranium, and cobalt.

One biotech company, *The Institute for Genomic Research*, has successfully sequenced a microbe that can absorb large amounts of radioactivity. Company scientists hope to use the genes that code for the "uranium-gobbling pathway" to fashion new biological means of cleaning up radioactive dump sites. With more than 200 million tons of hazardous materials being generated annually in the U.S. alone, and the costs of cleaning up toxic waste sites now estimated to be in excess of $1.7 trillion, industry analysts see bioremediation as one of the growth industries in the Biotech Century.

In agriculture, bioengineering is being looked to as a partial substitute for petrochemical farming. Scientists are busy at work engineering new food crops that can take in nitrogen directly from the air, rather than having to rely on the more costly petrochemical-based fertilizers currently in use. There are also experiments under way to transfer desirable genetic characteristics from one species to another in order to improve the nutritional value of plants and increase their yield and performance. Scientists are experimenting with genes that confer resistance to herbicides, help ward off viruses and pests, and can adapt a plant to salty or dry terrains, all in an effort to upgrade and speed the flow of agricultural products to market.

The first commercially grown gene-spliced food crops were planted in 1996. More than three-quarters of Alabama's cotton crop was genetically engineered to kill insects. In 1997, farmers planted genetically engineered soy on more than 8 million acres and genetically engineered corn on more than 3.5 million acres in the United States. The chemical and agribusiness companies hope to see a majority of farmlands converted over to gene-spliced crops within the next five years.

Meanwhile, the first genetically engineered insect, a predator mite, was released in Florida in 1996. Researchers at the University of Florida hope it will eat other mites that damage strawberries and other crops. Scientists at the University of California at Riverside are inserting a lethal gene into the pink bollworm, a caterpillar that causes millions of dollars of damage to the nation's cotton fields each year. The killer gene becomes activated in the offspring, killing young caterpillars before they can damage the cotton, mate, and reproduce. Researchers Thomas Miller and John Peloquin hope to raise millions of the genetically engineered bollworms to adulthood and then release them into the environment to mate with wild bollworm moths. The offspring will contain the lethal gene and die en masse in this new form of pest management.

Several biotech companies are working in the new field of tissue culture research, with the goal of moving more agricultural production indoors in the coming century. In the late 1980s, a U.S.-based biotechnology firm, Escagenetics of San Carlos, California, announced it had successfully produced vanilla from plant-cell cultures in the laboratory. Vanilla is the most popular flavor in America. One-third of all ice cream sold in America is vanilla. Vanilla is expensive to produce, however. The plant has to be hand-pollinated and requires special attention in the harvesting and curing processes.

Now, the new gene-splicing technologies allow researchers to produce commercial volumes of vanilla in laboratory vats, by isolating the gene that encodes the metabolic pathway that yields vanilla flavor and growing it in a bacteria bath, eliminating the bean, the plant, the soil, the cultivation, the harvest, and the farmer. Researchers have also successfully grown orange and lemon vesicles from tissue culture, and some industry analysts believe that the day is not far off when orange juice will be "grown" in vats, eliminating the need for planting orange groves.

The many changes taking place in agriculture are being accompanied by revolutionary changes in the field of animal husbandry. Researchers are developing genetically engineered "super animals" with enhanced characteristics for food production. They are also creating novel transgenic animals to serve as "chemical factories" to produce drugs and medicines and as organ "donors" for human transplants.

Much of the cutting edge research in animal husbandry is occurring in the new field of "pharming." Researchers are transforming herds and flocks into bio-factories to produce pharmaceutical products, medicines, and nutrients. The new pharming technology moved a step closer to commercial reality on February 22, 1997, when Ian Wilmut, a fifty-two-year-old Scottish embryologist, announced the cloning of the first mammal in history—a sheep named Dolly. Wilmut replaced the DNA in a normal sheep egg with the DNA from the mammary gland of an adult sheep. He tricked the egg into growing and inserted it into the womb of another sheep. The birth of Dolly is a milestone event of the emerging Biotechnological Age. It is now possible to mass-produce identical copies of a mammal, each indistinguishable from the original.

Researchers added a human gene to fetal sheep cells growing in a laboratory dish and then cloned a sheep from the enhanced cells. The experiment caught even normally staid scientists by surprise. Together, genetic manipulation and cloning will allow scientists to both customize and mass-produce animals, using the kind of quantifiable standards of measurement, predictability, and efficiency, that have heretofore been used to transform inanimate matter and energy into material goods.

Agribusiness, pharmaceutical, and chemical companies plan to mass-produce customized and cloned animals for use as chemical factories, to secrete a range of drugs and medicines. The meat industry is also interested in cloning. Being able to reproduce animals with exacting standards of lean-to-fat ratios and other features provides a form of strict quality control that has eluded the industry in the past. Animal clones will also be used to harvest organs for human transplantation. Being able to mass-produce exact replicas of animals will assure the kind of bioindustrial quality control that will be necessary to make xenotransplants a major commercial business in the Biotech Century.

Marine biotechnology is also expected to reap large profits in the coming decade. Most marine biotech research is geared to engineer customized fish that can be mass-produced through cloning techniques and be reared in fish farms. With one out of every five fish sent to market today coming from a fish farm, scientists hope the so-called "blue revolution" in marine biotechnology will rival the "green revolution" in agriculture.

Millions of people are already using genetically engineered drugs and medicines to treat heart disease, cancer, AIDS, and strokes. In 1995 researchers tested more than 284 new gene-spliced medicines, an increase of 20 percent over the previous year. Many conventional drugs have been replaced altogether with the new gene-spliced substitutes. Genetically engineered human insulin has virtually eliminated the use of naturally derived insulin from cows and pigs for more than 3.4 million Americans suffering from diabetes.

Erythropoietin, produced by Amgen, is used by nearly 200,000 people who are on kidney dialysis each year. The gene-spliced product stimulates the growth of red blood cells, reducing the need for risky blood transfusions. Genentech's tissue plasminogen activator (tPA) dissolves blood clots. Avonex and Betaseron, the beta-interferons, are used as therapies for multiple sclerosis. Pulmozyme (DNAase) is used to treat lung congestion in cystic fibrosis patients.

These new genetically engineered drugs are only the beginning of the vast possibilities that lie ahead, say researchers in the field. Scientists in a number of genetic engineering laboratories are working on new ways of altering the genetic characteristics of insects who are the carriers of deadly human diseases, rendering them harmless as infective agents. In one set of experiments, scientists have genetically engineered mosquitos with altered salivary glands making the insect unable to inject malarial parasites when it bites its victims. The genetically altered bacteria produce an antibiotic that kills the disease-causing parasite in the insect's digestive tract.

References

1. Cox, Earl D. and George S Paul (1996) *Beyond Humanity: Cyber Evolution and Future Minds*. Roackland, MA: Charles River Media.
2. *Dixon D (1981) After Man-A Zoology of the Future*. New York: St. Martin's Press.
3. Dodds JH (1985) *Plant Genetic Engineering*. New York: Cambridge University Press.

4. Goodman DC (1987) *From Farming to Biotechnology: A Theory of Agro-industrial Development*. Oxford: Blackwell.
5. Moravec, H (1999) *Robot: Mere Machine to Transcendant Mind*. New York: Oxford University Press.
6. Morholt E, Brandwein PF (1986) *Sourcebook for the Biological Sciences* 3rd Edition. San Diego: Harcourt Brace JOvanovich.
7. Postman, N (1992) *Technopoly: The Surrender of Culture to Technology*. New York: Vintage Books.
8. Retzer WJ (1991) Biotechnology Workbook. Englewood Cliffs, NJ: Prentice Hall.
9. Rifkin J (1998) The Biotech Century. Penguin, New York.
10. Woodrow Wilson National Fellowship Foundation (1992) Bioethics: 1992 Biology Institute. Princeton: Woodrow Wilson National Fellowship Foundation.

Glossary

agrobacterium: a naturally occurring bacterium that has the ability to transfer genes into plants such as tobacco and soybean.

allele: one of a number of different forms of a gene. Each person inherits two alleles for each gene, one allele from each parent. These alleles may be the same or may be different from one another.

amino acid: one of twenty different molecules that combine to form proteins. The sequence of amino acids in a protein determines the protein structure and function.

antibiotic: a class of natural and synthetic compounds that inhibit the growth of or kill other microorganisms.

antibiotic resistance: the ability of a microorganism to produce a protein that disables an antibiotic or prevents transport of the antibiotic into the cell.

antibody: a protein produced by the immune system in humans and higher animals, which binds to a specific antigen. When antibodies bind to corresponding antigens they set in motion a process to eliminate the antigens.

antigen: a foreign substance that, when introduced into the body, can stimulate an immune response.

aquaculture: the cultivation of plants using water as the support medium. It can also mean rearing marine life under controlled conditions in water, such as fish farming.

autoimmune disease: a disease whereby an individual's immune system mounts an attack on a portion of its own tissues. Tissues undergoing such an attack can be destroyed in the process. Rheumatoid arthritis is an example of an autoimmune disease.

B

***Bacillus thuringiensis (Bt)*:** a bacterium that kills insects; a major component of the microbial pesticide industry.

bacteriophage: a virus that infects bacteria. In genetic engineering, it is used to introduce genes into bacteria cells.

bacterium: a structurally simple single cell with no nucleus. One bacterium, Bacillus thuringiensis, produces a protein toxic to certain destructive insects. Having plants produce this protein after genetic engineering is a form of built-in pest-resistant in plants.

base: one of the molecules - adenine, guanine, cytosine, thymine, or uracil - which form part of the structure of DNA and RNA molecules. The order of bases in a DNA molecule determines the structure of proteins encoded by that DNA.

base pair (bp): two complementary nucleotide bases joined together by chemical bonds. The two strands of the DNA molecule are held together in the shape of a double helix by the bonds between base pairs. The base adenine pairs with thymine, and guanine pairs with cytosine.

biodegradable: able to be broken down by micro-organisms in the environment

biodiversity: the wide diversity and interrelatedness of earth organisms based on genetic and environmental factors, the variety of flora and fauna in the natural environment

bioethics: a discipline dealing with the ethical implications of biological research and applications.

biogas: methane that is produced from an anaerobic digestor

bioinformatics: the science of informatics as applied to biological research. Informatics is the management and analysis of data using advanced computing techniques. Bioinformatics is particularly important as an adjunct to genomics research, because of the large amount of complex data this research generates.

bioinsecticides: microorganisms and viruses acting as diseases to insects

biomass: the sum total of living organisms in a culture. It can be used as a source of fuel, energy, commodity chemicals, animal feed, and specialty products (e.g., flavours, fragrances, pigments).

biomaterials: materials with novel chemical, physical, mechanical, or "intelligent" properties, produced through processes that employ or mimic biological phenomena

bioremediation: the use of living systems or their products to degrade wastes into less toxic or non-toxic products

biosensor: an electronic device that uses biological molecules to detect various compounds. (Coupled system).

biotechnology: the application of biological research techniques to the development of products which improve human health, animal health, and agriculture.

***Bt* crops**: genetically engineered to carry the gene from the soil bacterium Bacillus thuringiensis. The bacteria produces a protein that is toxic when ingested by certain lepidopteran insects. Crops containing the Bt gene are able to produce this toxin, thereby providing protection throughout the entire plant.

***Bt* cotton:** genetically engineered to control tobacco budworms, bollworms, and pink bollworms.

***Bt* corn** (maize): genetically engineered to provide protection against the European corn borer.

buffer zone (or 'separation distance'): a strip of land separating different types of crop, wide enough to minimise the level of cross-pollination between them.

C

cell: the smallest structural unit of all living organisms that is able to grow and reproduce independently. A cell is a mass of living material, surrounded by a membrane.

cell culture/tissue culture: growth of cells or tissues in the laboratory or in vitro.

cell line: cells that grow and replicate continuously outside the living organism.

chloroplast: structure in plant cells in which photosynthesis occurs

chromosome: the DNA in a cell is divided into structures called chromosomes. Chromosomes are large enough to be seen under a microscope. In humans, all cells other than germ cells usually contain 46 chromosomes: 22 pairs of autosomes and either a pair of X chromosomes (in females) or an X chromosome and a Y chromosome (in males). In each pair of chromosomes, one chromosome is inherited from an individual's father and one from his or her mother.

chymosin: one of the enzymes which causes milk to clot and separate into curds and whey during cheese-making.

clone: a term which is applied to genes, cells, or entire organisms which are derived from - and are genetically identical to-a single common ancestor gene, cell, or organism, respectively. Cloning of genes and cells to create many copies in the laboratory is a common procedure essential for biomedical research. Note that several processes which are commonly described as cell "cloning" give rise to cells which are almost but not completely genetically identical to the ancestor cell. "Cloning" of organisms from embryonic cells occurs naturally in nature (e.g. with the occurrence of identical twins). The laboratory cloning of a sheep using the genetic material from a cell of an adult animal has recently been reported.

cloning: the use of techniques of molecular biology to produce multiple copies of segments of DNA, usually genes

cloning vector: DNA molecule originating from a virus, a plasmid, or the cell of a higher organism into which another DNA fragment of appropriate size can be integrated without loss of the vectors capacity for self-replication; vectors introduce foreign DNA into host cells, where it can be reproduced in large quantities. Examples are plasmids, cosmids, and yeast artificial chromosomes; vectors are often recombinant molecules containing DNA sequences from several sources.

code: the sequence of DNA-bases which forms the instructions for a given characteristic or trait.

copy gene: genetic material incorporating the genetic code for a desirable trait which has been copied from the DNA of the donor to the host organism. (It is not technically possible to take a gene from a donor organism and insert it directly into the host organism).

crop rotation: a farming technique whereby different types of crops are grown over successive seasons.

cross-pollination: transfer of pollen from a flower of one plant or population to the stigma of a flower of another plant or population.

cultivar: a variety of plant produced through selective breeding by humans and maintained by cultivation.

cytoplasm: the material inside a cell located between the cell's nucleus and the cell membrane. Cytoplasm is a mixture of water, dissolved ions, sugars, and proteins. Cytoplasm includes a number of filaments and tubules, all arranged in a three-dimensional lattice, that provide support for the cell's organelles.

D

diploid cell: cell containing two structurally identical sets of chromosomes, except sex chromosomes

DNA (deoxyribonucleic acid): the molecule that encodes genetic information. DNA is a double-stranded helix held together by bonds between pairs of nucleotides.

DNA bank: a service which stores DNA extracted from blood samples or other human tissue. DNA samples stored in the DNA bank may be used, for example, to help individuals and their physicians to trace the pattern of disease in families, or for the purposes of future medical research.

DNA fingerprint: profile of an organism's genetic material, typically determined from DNA segments, in order to illuminate the differences between and among individuals

DNA polymerase: enzyme catalysing DNA synthesis from deoxyribonucleotides and a template DNA molecule

DNA polymorphism: multiple alternative sequences of DNA occurring at a particular gene site

DNA probe: fragment of DNA used to recognize a specific complementary DNA sequence

DNA sequencing: technique for deciphering the order of nucleotide bases in a DNA segment

DNA probe: a piece of single-stranded DNA, typically labelled so that it can be detected (for example, a radioactive or fluorescent label can be used), which can single out and bind with (and only with) another specific piece of DNA. DNA probes can be used to determine which sequences are present in a given length of DNA or which genes are present in a sample of DNA.

DNA repair genes: genes which code for proteins which correct mistakes in DNA sequences. When these genes are altered, mutations may be able to accumulate in the genome, ultimately resulting in disease.

double helix: describes the coiling of the antiparallel strands of the DNA molecule, resembling a spiral staircase in which the paired bases form the steps and the sugar-phosphate backbones form the rails.

E

E. coli (Escherichia coli) **:** common bacterium that has been studied intensively by geneticists because of its small genome size, normal lack of pathogenicity, and ease of growth in the laboratory.

ecology: the study of the interactions of organisms with their environment and with each other.

ecosystem: the organisms in a plant population and the biotic and abiotic factors which impact on them.

electrophoresis: a method of separating large molecules (such as DNA fragments or proteins) from a mixture of similar molecules. An electric current is passed through a medium containing the mixture, and each kind of molecule travels through the medium at a different rate, depending on its electrical charge and size. Separation is based on these differences. Agarose and acrylamide gels are the media commonly used for electrophoresis of proteins and nucleic acids.

embryo: the stage of an organism's development directly following fertilization and zygote formation. The term embryo is often used to describe all developmental phases between the first cell cleavage and birth/hatching of a developing organism.

enzyme: a protein that facilitates a biochemical reaction in a cell. In general, these biochemical reactions would not occur if the enzyme is not present. For example, an enzyme can facilitate (also called catalyze) the destruction of another protein by breaking the bonds between amino acids. An enzyme of that type is called a protease.

ES cells: embryonic stem cells. Cultured cells derived from the pluripotent inner cell mass of blastocyst stage embryos. Used for gene targeting by homologous recombination.

eugenics: a field of study, which gained some prominence at the beginning of the 20th century, which advocated the manipulation of human characteristics by artificial selection.

eukaryote: cell or organism with membrane-bound, structurally discrete nucleus and other well-developed subcellular compartments. Eukaryotes include all organisms except viruses, bacteria, and blue-green algae. Compare prokaryote.

***ex vivo*:** [Latin] pertaining to a biological process or reaction taking place outside of a living cell or organism.
exogenous DNA: DNA which has been introduced into an organism but which originated outside that organism (e.g. material inserted into a cell by a virus).
expressed sequence tag (EST): a short strand of DNA (approximately 200 base pairs long) which is part of a cDNA. Because an EST is usually unique to a particular cDNA, and because cDNAs correspond to a particular gene in the genome, ESTs can be used to help identify unknown genes and to map their position in the genome.

F

full gene sequence: the complete order of bases in a gene. This order determines which protein a gene will produce.
functional food: food enriched or fortified with components or ingredients offering a specific health benefit other than a strict nutritional effect
fungicide: a substance that kills fungi
fungus: type of microorganism with distinct nucleus and cytoplasm, e.g. bread mould.
fusion: The process of joining the membranes of two cells to create another cell that contains the nuclear material from both parent cells.

G

gamete: mature reproductive cell, usually haploid, serving fertilization
gene: a length of DNA which codes for a particular protein, or in certain cases a functional or structural RNA molecule.
gene candidate: is designated when a DNA has homology to something of known function but no product was produced to show it is the correct site.
gene expression: the process by which the information in a gene is used to create proteins.
gene families: groups of closely related genes that make similar products
gene flow: gradual exchange of genetic material between populations caused by the dispersal of gametes or the migration of individuals

gene library: a collection of cloned DNA fragments which, taken together, represent the entire genome of a specific organism. Such libraries or "gene banks" are assembled so as to allow the isolation and study of individual genes. Gene libraries are produced by first breaking up or fractionating an entire genome. This fractionation can be accomplished either by physical methods or by use of restriction enzymes. The genome fragments are then cloned (multiplied in number) and stored for later use.

gene pool: the total genetic material of a freely interbreeding population at a given time

gene product: the protein produced by a gene.

gene targeting: knocking out the functioning of specific genes of an organism by genetic manipulation

gene therapy: Gene therapy of either kind should be clearly distinguished from the use of genomics to discover new targets for drug discovery and new diagnostic tools.

genetic code: the set of codons in DNA or mRNA. Each codon is made up of three nucleotides which call for a unique amino acid. For example, the set AUG (adenine, uracil, guanine) calls for the amino acid methionine. The sequence of codons along an mRNA molecule specifies the sequence of amino acids in a particular protein.

genetic counseling: analysis of risk for genetic defects in a family and presentation of options to avoid or ameliorate possible risks.

genetic disease: a disease that has its origin in changes to the genetic material, DNA. Usually refers to diseases that are inherited in a Mendelian fashion, although non-inherited forms of cancer also result from DNA mutation.

genetic engineering: altering the genetic material of cells or organisms in order to make them capable of making new substances or performing new functions.

genetic line: A variety or strain known to possess valuable characters.

genetic map: a map of a genome which shows the relative positions of the genes and/or markers on the chromosomes.

genetic marker: see marker.

genetic model: the overall specification of how the disease allele(s) act to influence the disease. For parametric (model-dependent) linkage analysis, the genetic model must be specified for the analysis. Components of the genetic model include information on whether the disorder is autosomal or X-linked, dominant or recessive, the frequency and penetrance of the disease allele, the frequency of phenocopies, and the mutation rate.

genetic modification: a technique whereby individual genes can be copied and transferred to another living organism to alter its genetic make up and thus incorporate or delete specific characteristics into or from the organism.

genetic mosaic: an organism which contains cells with different genetic information. This can result from a mutation occurring in a cell (or cells) at some point during development, or by the fusion of embryos at an early stage in their respective development. Such organisms develop into adult organisms that are also referred to as "chimeras." Organisms that express both male and female characteristics by virtue of being a mosaic of male and female cells are referred to as "gyandromorphs".

genetic mutation: a change in the nucleotide sequence of a DNA molecule. Genetic mutations are a kind of genetic polymorphism. The term "mutation," as opposed to "polymorphism," is generally used to refer to changes in DNA sequence which are not present in most individuals of a species and either have been associated with disease (or risk of disease) or have resulted from damage inflicted by external agents (such as viruses or radiation).

genetic polymorphism: a difference in DNA sequence among individuals, groups, or populations (e.g. a genetic polymorphism might give rise to blue eyes versus brown eyes, or straight hair versus curly hair). Genetic polymorphisms may be the result of chance processes, or may have been induced by external agents (such as viruses or radiation). If a difference in DNA sequence among individuals has been shown to be associated with disease, it will usually be called a genetic mutation. Changes in DNA sequence which have been confirmed to be caused by external agents are also generally called "mutations" rather than "polymorphisms."

genetic predisposition: susceptibility to a disease which is related to a genetic mutation, which may or may not result in actual development of the disease.

genetic testing: the analysis of an individual's genetic material. Among the purposes of genetic testing could be to gather information on an individual's genetic predisposition to particular health conditions, or to confirm a diagnosis of genetic disease.

genomic DNA: the basic chromosome set consisting of a species-specific number of linkage groups and the genes contained therein.

genome: all the genetic material in the chromosomes of a particular organism; its size is generally given as its total number of base pairs.

genomic healthcare: health care which utilizes advances made by the science of genomics.

genomic library: a collection of clones made from a set of randomly generated overlapping DNA fragments representing the entire genome of an organism. Compare library, arrayed library.

genomics: the study of genes and their function. Recent advances in genomics are bringing about a revolution in our understanding of the molecular mechanisms of disease, including the complex interplay of genetic and environmental factors. Genomics is also stimulating the discovery of breakthrough healthcare products by revealing thousands of new biological targets for the development of drugs, and by giving scientists innovative ways to design new drugs, vaccines and DNA diagnostics. Genomics-based therapeutics include 'traditional' small chemical drugs, protein drugs, and potentially gene therapy.

genotype: the particular genetic pattern seen in the DNA of an individual. "Genotype" is usually used to refer to the particular pair of alleles that an individual possesses at a certain location in the genome.

germ cell: reproductive cell: sperm and egg cells, and their precursors. Germ cells have only one set of chromosomes (23 in all), while all other cells have two copies (46 in all).

germ line gene therapy: germ line gene therapy involves the insertion of normal genes into germ cells or fertilized eggs in an attempt to create a beneficial genetic change which can be transmitted to an organism's offspring (for example, to correct for a genetic trait associated with disease). If a change is introduced via germ line gene therapy, that change may be present in the offspring from birth in every cell in the body.

GM, GMO: see genetic modification

golden rice: a strain of genetically altered rice to combat vitamin A deficiency, containing three transplanted genes that allow plants to produce rice kernels containing beta-carotene, a compound that is converted to vitamin A in the human body. Vitamin A deficiency is the world's leading cause of blindness and a malaise that affects as many as 250 million children.

green Revolution: advances in genetics, petrochemicals, and machinery that culminated in a dramatic increase in crop productivity during the third quarter of the 20th century.

H

haploid cell: cell containing only one set (or half the usual diploid number) of chromosomes.

herbicide: any substance that is toxic to plants; usually used to kill specific unwanted plants.

human Genome Project: an international research effort aimed at discovering the full sequence of bases in the human genome. Led in the United States by the National Institutes of Health and the Department of Energy.

hybrid: the progeny of genetically dissimilar parents; a heterozygote.

hybridisation: the process of joining two complementary strands of DNA or one each of DNA and RNA to form a double-stranded molecule.

I

identical twin: twins which have been produced by the division of a single zygote (monozygotic). Each twin has an identical genotype. While each twin begins with the same set of genetic information, the effect of the environment within which each individual grows up can cause differences in how their genetic make-up is expressed.

immune response: a physiological response produced in humans and higher animals, to defend the body against the introduction of foreign material. "Hay fever," for example, is the response sometimes produced by the human immune system to the inhalation of pollen.

immune system: a biological defense system that has evolved in vertebrates to protect them against the introduction of foreign material (such as pollen, or invading micro-organisms) and to prevent the body from developing cancer. See also immune response and antigen.

immunodeficient : a condition wherein an organism's immune system has been compromised and becomes defective either as the result of an inherited genetic error, by the actions of an infectious agent, as the result of disease, or as the result of having been exposed to immunosuppressant drugs such as are used after transplant surgery to prevent rejection.

immunosuppressant: a compound which reduces or halts an immune response.

immunosuppression : a reduction or elimination of an immune response.

imprinting: a phenomenon in which the phenotype of the disease depends on which parent passed on the disease gene. For instance, Prader-Willi syndrome and Angelman syndrome are both inherited when the same part of chromosome 15 is missing. When the father's complement of 15 is missing, then the child has Prader-Willi, but when the mother's complement of 15 is missing, the child has Angelman syndrome.

insect resistant: a plant that contains substances (proteins) which kill or deter plant-eating insects.

insecticide: a substance that kills insects

***in situ* hybridisation:** use of a DNA or RNA probe to detect the presence of the complementary DNA sequence in cloned bacterial or cultured eukaryotic cells.

insulin: a peptide hormone secreted by the islets of Langerhans of the pancreas that regulates the level of sugar in the blood.

integrated Crop Management: farming techniques combining pest management, crop rotation and other methods of production in a way that is both economically viable and environmentally responsible.

interferon: a family of small proteins that stimulate viral resistance in cells.

intergenic regions: DNA sequences located between genes that comprise a large percentage of the human genome with no known function.

***in vitro*:** pertaining to a biochemical process or reaction taking place in a test-tube (or more broadly, in a lab) as opposed to taking place in a living cell or organism.
***in vitro* fertilization (IVF):** fertilization of an egg cell by sperm cells performed "in glass", (i.e. in a laboratory procedure. A pre-embryo is formed as the result of this process. This process is often used to achieve pregnancy if the mother's fallopian tubes are damaged or missing.
***in vitro* maturation (IVM) :** a laboratory process whereby a primary oocyte obtained from an immature tertiary follicle is allowed to mature until it becomes a secondary oocyte and capable of being fertilised by a sperm cell. Fertilisation is then performed in the laboratory by use of using in vitro fertilisation. This process is often used to increase the number of egg cells available for IVF so as to increase the chances of successful fertilisation, implantation, and pregnancy.
***in vivo*:** pertaining to a biological process or reaction taking place in a living cell or organism.
***in vivo* fertilization (IVF) :** fertilisation which occurs within the fallopian tubes.
***in vivo* gene therapy:** gene repair or alteration performed within individual organism. Examples include the insertion of genes into lung cells of patients suffering from cystic fibrosis.
intercross: a cross between two animals that have the same heterozygous genotype at designated loci; for example, between sibling F1 hybrids that were derived from an outcross between two inbred strains.
interphase: the period in the cell cycle when DNA is replicated in the nucleus; followed by mitosis.

K

karyotype: a photomicrograph of an individuals chromosomes arranged in a standard format showing the number, size, and shape of each chromosome type; used in low-resolution physical mapping to correlate gross chromosomal abnormalities with the characteristics of specific diseases.

L

library: a set of clones of DNA sequences from an organism's genome. A particular library might include, for example, clones of all of the DNA sequences expressed in a certain kind of cell, or in a certain organ of the body

linkage: refers to the tendency of certain genes to be inherited together. Two genes are said to be "linked" if they are often inherited together, due to their close proximity on a chromosome.

linkage map: a map of the relative positions of genetic loci on a chromosome, determined on the basis of how often the loci are inherited together. Distance is measured in centimorgans (cM).

lipids: organic substances (e.g. fats, oils, steroids etc.), insoluble in water and soluble in organic solvents.

living modified organism: a living organism that has a new combination of genetic material obtained through the use of modern biotechnology.

localize: determination of the original position (locus) of a gene or other marker on a chromosome.

locus (pl. loci): the position on a chromosome of a gene or other chromosome marker; also, the DNA at that position. The use of locus is sometimes restricted to mean regions of DNA that are expressed.

lymphocyte: a white blood cell that is important in the body's immune response, because they contain nuclei, lymphocytes serve as an easily accessible source for genomic DNA.

M

map distances: maps of the human genome are generally of two types, genetic and physical. Genetic maps measure the amount of recombination between two loci and genetic map units are quantified as either % recombination (theta) or cM between two loci. Physical maps aim to quantify the actual amount of DNA, usually in base pairs, between two loci. 1% recombination is equivalent to 1,000,000 base pairs of DNA.

mapping: see genetic map

mapping population: a group of organisms sharing a common gene pool used in the construction of a genetic map.

marker: a sequence of bases at a unique physical location in the genome, which varies sufficiently between individuals that its pattern of inheritance can be tracked through families and/ or it can be used to distinguish among cell types. A marker may or may not be part of a gene. Markers are essential for use in linkage studies and genetic maps to help scientists to narrow down the possible location of new genes, and to discover the associations between genetic mutations and disease.

meiosis: the reduction division process by which haploid gametes and spores are formed, consisting of a single duplication of the genetic material followed by two mitotic divisions.

Mendelian inheritance: the mode of inheritance of nuclear genes, in contrast to "non-Mendelian" or "extra-nuclear" inheritance which is controlled by cytoplasmic heredity determinants.

microorganism: an organism that can be seen only under a microscope. Categories of microorganisms include Algae, Bacteria, Fungi, Protozoa, Viruses, or Subviral Agents. Also referred to as microbe.

mitochondrial DNA: (mtDNA) : DNA that is found within the mitochondria of all eukaryotic organisms. Since this DNA is located outside of a cell's nucleus it is referred to as extranuclear DNA. Mitochondrial DNA is found in the form of a circular chromosome which has much in common with that found within prokaryotic organisms (bacteria). Mitochondrial DNA is self-replicating (i.e. it replicates independent of the DNA found in a cell's nucleus). Mitochondria are inherited by an organism from the large female (maternal) egg cell, not from the much smaller male sperm cell. As such, mitochondrial DNA can be studied to understand an organism's maternal lineage, one unaffected by the genetic recombination that occurs during sexual reproduction. Since changes in the mitochondrial genome are not affected by repeated and frequent sexual recombination, the rate at which mtDNA changes occur can be exploited in understanding the evolution of a particular species over long periods of time. Damage to mitochondrial DNA can result in disease.

mitochondrial genetic disorder: a mutation in the mitochondrial genome can result in diseases which affect an entire organism. Since mitochondrial DNA is inherited from the maternal parent, not the paternal parent, mitochondrial genetic disorders are inherited from the female parent. This happens by virtue of the fact that zygotes in organisms such as mammals inherit their mitochondria from the large maternal egg cell, not the small paternal sperm cell. Note that some mitochondrial· disorders can result from damage caused by therapeutic drugs or the ageing process.

mitochondrion: an organelle within a cell that generates most of the cell's energy. Its DNA also maintains and expresses genetic information.

mitosis: the process of nuclear division in cells that produces daughter cells that are genetically identical to each other and to the parent cell.

molecular biology: the study of the biochemical and molecular interactions within living cells.

molecular cloning: the biological amplification of a specific DNA sequence through mitotic division of a host cell into which it has been transformed or transfected.

monogenic disorder: a disorder which is caused by mutations in just one gene (such as hemophilia), as opposed to polygenic disorders (such as hypertension) which involve the interaction of several genes. Monogenic disorders are also known as "single-gene disorders," or as "Mendelian disorders" because their inheritance patterns tend to follow the genetic laws first described by Gregor Mendel.

monogenic inheritance: the inheritance of character differences controlled by the alleles for one particular locus, as opposed to di-,tri-, or polygenic control exerted by two, three, or many nonallelic genes.

monozygotic twins : twins which have been produced by the division of a single zygote. Each twin has an identical genotype. As such, these individuals are often referred to as "identical twins". While each twin begins with the same set of genetic information, the effect of the environment within which each individual grows up can cause differences in how their genetic make-up is expressed.

multifactorial: a trait is considered to be multifactorial in origin when two or more genes, together with an environmental effect, work together to lead to a phenotype.

multiple gene line: a mixture of pure lines or cultivated varieties that are phenotypically similar in all important agronomic characteristics but are genotypically different.

multiple tolerance (to herbicides): of a plant, being able to tolerate anumber of different herbicides.

multiplexing: a sequencing approach that uses several pooled samples simultaneously, greatly increasing sequencing speed.
mutagenesis: a process by which the genetic information of an organism is changed in a stable, heritable manner, either in nature or experimentally by the use of chemicals or radiation. In agriculture, these genetic changes are used to develop useful traits.
mutation: a change, deletion, or rearrangement in the DNA sequence that may lead to the synthesis of an altered inactive protein the loss of the ability to produce the protein. If a mutation occurs in a germ cell, then it is a heritable change in that it can be transmitted from generation to generation. Mutations may also be in somatic cells and are not heritable in the traditional sense of the word, but are transmitted to all daughter cells.

N

narrow spectrum herbicides: herbicides that kill a limited range of plant types.
nematode: a roundworm or threadworm that attacks plants and animals.
nitrogen fixation: property of microorganisms to convert nitrogen gas from the atmosphere into a chemically combined form, ammonia, which is essential to plant growth.
novel trait in a plant: a plant possessing characteristics not normally found in that species where the new characteristic has been created through specific genetic manipulation, transformation, mutation, etc.
nuclear chromosome: one of the threadlike structures consisting of chromatin and carrying genetic information arranged in a linear sequence, found in the nucleus of eukaryotic cells.
nuclear transfer: the transfer of a diploid cell nucleus from a donor cell into another cell that has had its original nucleus removed (enucleated cell).
nuclease: enzyme breaking bonds (hydrolysing) in nucleic acid molecules.
nucleic acid: one of the family of molecules which includes the DNA and RNA molecules. Nucleic acids were so named

because they were originally discovered within the nucleus of cells, but they have since been found to exist outside the nucleus as well.

nucleotide: the "building block" of nucleic acids, such as the DNA molecule. A nucleotide consists of one of four bases - adenine, guanine, cytosine, or thymine - attached to a phosphate-sugar group. In DNA the sugar group is deoxyribose, while in RNA (a DNA-related molecule which helps to translate genetic information into proteins), the sugar group is ribose, and the base uracil substitutes for thymine. Each group of three nucleotides in a gene is known as a codon. A nucleic acid is a long chain of nucleotides joined together, and therefore is sometimes referred to as a "polynucleotide."

nucleus: the membrane bound structure containing a cell's central DNA found within all eukarotic cells.

O

oncogene: a gene which is associated with the development of cancer.

oocyte: cells produced within an ovary during the process of producing egg cells. Primary oocytes are large cells produced in great number by mitosis. Meiosis occurs within primary oocytes to produce two cells: one is a small primary polar body, the other is a large secondary oocyte (containing most of the original primary oocyte's cytoplasm). The secondary oocyte undergoes a second meiotic division to produce a large ovum (again, with most of the cytoplasm) and a small secondary polar body. The ovum or egg cell is the female gamete that participates in the fertilization process.

orphan drug products: drugs, vaccines or in vivo diagnostic agents which physicians use to treat, prevent or diagnose rare diseases. Usually, such drugs are not commercially viable, the financial return being small compared with R&D and marketing costs. To help overcome this problem the EU introduced an Orphan Drugs regulation in spring 2000.

P

pathogen: specific causative agent of disease

pathogenic microorganism: microorganism causing disease to animals and plants

parental line: The breeding line from which a parent of a cross is taken.

parthenogenesis: The development of an organisms from an unfertilized egg. This process isrelatively common in plants, but less so among animals. Some species of insects can produce large numbers of individuals which are diploid. Some species of lizards can also produce diploid progeny. Since a female parent is, in essence, cloning herself, parthenogenesis always produces only female offspring. Since no recombination of genetic material between parents occurs, genetic variation within populations that use parthenogenesis is limited. However, this process can result in quick repopulation of a devastated area in a short period of time especially if the population has been thinned such that mating encounters between males and females are rare.

patent: a grant issued by the government, which gives the patent holder the right to exclude others from making, using, or selling a patented invention for a certain term. In most countries, the term begins on the date on which the patent issues ends and 20 years from the date on which the application for the patent was filed. Patents are granted on inventions which meet the requirements of novelty, non-obviousness, and utility. A patent holder cannot use a patented invention dominated by the patent of another, absent a license or cross-license.

PCR: see polymerase chain reaction

peptide: two or more amino acids chained together by a bond called a "peptide bond." A protein is a long chain of amino acids joined together in this way, and therefore is sometimes referred to as a polypeptide. Some proteins contain more than one polypeptide chain.

pesticide: a substance that kills harmful organisms (for example, an insecticide or fungicide).

pharmacogenomics: the science of understanding the correlation between an individual patient's genetic make-up (genotype) and their response to drug treatment. Some drugs work well in some patient populations and not as well in others. Studying the genetic basis of patient response to therapeutics allows drug developers to more effectively design therapeutic treatments.

phenocopy: a trait which appears to be identical to a genetic trait, but which is caused by non-genetic factors.

phenotype: a set of observable physical characteristics of an individual organism. A single characteristic can be referred to as a "trait," although a single trait is sometimes also called a phenotype. For example, blond hair could be called a trait or a phenotype, as could obesity. A phenotype can be the result of many factors, including an individual's genotype, environment, and lifestyle, and the interactions among these factors: the observed manifestation of a genotype. The phenotype may be expressed physically, biochemically, or physiologically.

physical map: a map of the locations of identifiable landmarks on DNA (e.g., restriction enzyme cutting sites, genes), regardless of inheritance. Distance is measured in base pairs. For the human genome, the lowest-resolution physical map is the banding patterns on the 24 different chromosomes; the highest-resolution map would be the complete nucleotide sequence of the chromosomes.

plasmid: a structure composed of DNA that is separate from the cell's genome. In bacteria, plasmids confer a variety of traits and can be exchanged between individuals- even those of different species. Plasmids can be manipulated in the laboratory to deliver specific genetic sequences into a cell.

point: a single nucleotide on a chromosome.

polygenic: pertaining to a phenotype that results from interactions among the products of two or more genes with alternative alleles.

Polymerase Chain Reaction (PCR): a method for creating millions of copies of a particular segment of DNA. If a scientist needs to detect the presence of a very small amount of a particular DNA sequence, PCR can be used to amplify the amount of that sequence until there are enough copies available to be detected.

prokaryote: cell or organism lacking a membrane-bound, structurally discrete nucleus and other subcellular compartments. Bacteria are prokaryotes. Compare eukaryote.

protein: a biological molecule which consists of many amino acids chained together by peptide bonds. The sequence of amino acids in a protein is determined by the sequence of nucleotides in a DNA molecule. As the chain of amino acids is being synthesized, it is also folded into higher order structures shaped, for example, like helices or like flat sheets. Proteins are required for the structure, function, and regulation of cells, tissues, and organs in the body.

protein drug: a drug whose main constituent is a protein.

R

recombinant: the result of a crossover in a doubly heterozygous parent such that alleles at two loci that were present on opposite homologs are brought together on the same homolog. The term is used to describe the chromosome as well as the animal in which it is present.

recombination: recombinant DNA is produced when genetic information from more than one organism is recombined in a laboratory process into a hybrid molecule.

recombination map: the representation of the genetic distance separating nonallelic gene loci in a linkage structure or the arrangement of the mutational sites of a particular gene, by utilising the frequency of genetic recombination via crossing-over between genes and within genes.

refuge: in farming, an area of land near to Genetically Modified (GM) crops , kept free of GM plants where similar non-GM crops are grown, often without pesticides, to allow the development of a pest population.

regulatory gene: a gene which controls the protein-synthesizing activity of other genes.

resistance: the development of a characteristic within an organism (e.g. an insect) whereby it becomes able to protect itself from the effects of a particular substance (e.g. an insecticide)

resolution: degree of molecular detail on a physical map of DNA, ranging from low to high.

retrovirus: a virus that has RNA as its genetic material. When introduced into a host cell, it is used as a template to produce viral DNA, which leads to the formation of more, identical viruses.

ribonucleic acid (RNA): a molecule that translates the instructions encoded in DNA to build proteins.

ribosome: a part of a cell that builds proteins by linking amino acids according to the sequence on a strand of messenger RNA.

S

segregation: the principle that the two partners of a chromosome pair are separated during meiosis and distributed randomly to the germ cells. Each germ cell has an equal chance of receiving either chromosome.

selectable marker: a gene whose expression allows one to identify cells that have been transforrned or transfected with a vector containing the marker gene.

self-pollinating: transferring pollen from the stamens within a flower to the stigma of the same flower so that it produces seeds containing the same genetic information as the parent.

sequence: the order of nucleotides in a DNA or RNA molecule, or the order of amino acids in a protein.

sequence tagged site (STS): short (200 to 500 base pairs) DNA sequence that has a single occurrence in the human genome and whose location and base sequence are known. Detectable by polymerase chain reaction, STSs are useful for localizing and orienting the mapping and sequence data reported from many different laboratories and serve as landmarks on the developing physical map of the human genome. Expressed sequence tags (ESTs) are STSs derived from cDNAs.

sequencing: determining the order of nucleotides in a DNA or RNA molecule, or determining the order of amino acids in a protein.

sex chromosomes: the X and Y chromosomes in human beings that determine the sex of an individual. Females have two X chromosomes in diploid cells; males have an X and a Y chromosome. The sex chromosomes comprise the 23rd chromosome pair in a karyotype. Compare autosome.

sex-linked: a genetic trait whose genes are located on a sex chromosome (X or Y in humans).

shotgun method: cloning of DNA fragments randomly generated from a genome. See library, genomic library.

single-gene disorder: see monogenic disorder

somatic cell: a somatic cell is any cell of the body except for germ cells (sperm cells and egg cells) and their precursors.

somatic cell gene therapy: somatic cell gene therapy involves the insertion of genes into cells for therapeutic purposes, for example to induce the treated cells to produce a protein that the body is missing. It does not affect the genetic makeup of a patient's offspring, and generally does not change all, or even most, cells in the recipient. Somatic cell gene therapy is only one way of applying the science of genomics to improve health care. Others include the use of genomics to yield new targets for drug discovery, to refine the process of drug development and to discover new diagnostics. Compare germ line gene therapy.

somatic cell genetic mutation: a genetic mutation in a somatic cell. Such mutations, which are not inherited from parents but occur during the lifetime of an organism, are often known as "acquired genetic mutations." Somatic cell genetic mutations are not passed on to offspring. Compare germ line genetic mutation.

somatic cell nuclear transfer: a process whereby the nucleus of a somatic cell is removed and placed into an enucleated oocyte (i.e. an egg cell that has had its own nucleus i.e. all genetic information removed). The net result is to have the genetic information from the donor nucleus in control of the resulting cell. With further manipulation, such cells can be induced to form embryos. This process was used to create the cloned sheep "Dolly".

stem cell: undifferentiated, primitive cells in the bone marrow with the ability both to multiply and to differentiate into specific blood cells.

strain: a genetic type within an organism.

strain distribution pattern (SDP): the distribution of two segregating alleles at a single locus across a group of animal samples used for analysis in a linkage study. Used in the context of backcross data and data obtained from RI strains.

substantial equivalence: a principle inherent in the safety assessment process that compares a genetically modified food with a conventional non-modified food with a long history of safe use. If the modified food has essentially all the characteristics of the non-modified food with respect to food and feed value it is said to be substantially equivalent.

suppressor gene: a gene which helps to reverse the effects of damage to an individual's genetic material, typically effects which might lead to uncontrolled cell growth (as would occur in cancer). A suppressor gene may, for example, code for a protein which checks genes for misspellings, and/or which triggers a cell's self-destruction if too many genetic mutations have accumulated.

sustainable: of agriculture, whereby the farmer is able to increase or maintain crop yields long term, whilst conserving natural resources and protecting the environment.

T

technology transfer: the process of transferring discoveries made by basic research institutions to the commercial sector, to be developed into useful products and services.

tissue culture: a process where individual cells (or clumps) of plant or animal origin are grown artificially.

tolerance: the range of an environmental factor (also pathogens and pests) within which an organism or a population can survive.

totipotent: cells that have the ability to develop into any of the many different cell types which make up multicellular organisms. Embryos are composed of large numbers of totipotent cells which decline in number as development proceeds and cell specialization begins to occur. Adults have a much more limited ability to produce totipotent cells than embryos do. Organisms such as humans retain a complete set of genetic information in all adult body cells yet only a small fraction of an adult's cells have the ability to develop into multiple cell types. Recent research has shown that differentiated adult cells can be treated such that they become totipotent. Such totipotent or "stem cells" offer the possibility of a number of therapeutic uses such as repairing heart muscle after a heart attack or brain function after a stroke. Plant cells tend to retain much more of a capability of becoming totipotent - even in mature plants - than those of animals do.

toxin: a poison, usually originating in a plant or micro-organism

trait: a distinguishing characteristic or quality of an organism.

transcription: the process during which the information in a length of DNA is used to construct an mRNA molecule.

transfer RNA (tRNA): RNA molecules which bond with amino acids and transfer them to ribosomes, where protein synthesis is completed

transformation: a process by which the genetic material carried by an individual cell is altered by incorporation of exogenous DNA into its genome.

transgenic: an organism whose genome has been altered by the inclusion of foreign genetic material. This foreign genetic material may be derived from other individuals of the same species or from wholly different species. Genetic material may also be of an artificial nature. Foreign genetic information can be added to the organism during its early development and incorporated in cells of the entire organism. As an example, mice embryos have been given the gene for rat growth hormone allowing mice to grow into large adults. Genetic information can also be added later in development to selected portions of the organism. As an example, experimental genetic therapy to treat cystic fibrosis involves selective addition of genes responsible for lung function and is administered directly to the lung tissue of children and adults. Transgenic organisms have been produced that provide enhanced agricultural and pharmaceutical products. Insect resistant crops and cows that produce human hormones in their milk are just two examples.

V

vaccine: a preparation that contains an antigen made up of disease-causing organisms in a dead or weakened state. It is used to boost immunity against the given diseases, and can be created using the recombinant DNA process.
variety: a group of organisms within a species, having similar characteristics but not distinct enough to be a separate species.
vector: [1] an organism which serves to transfer a disease-causing organism (pathogen) from one organism to another. [2] a mechanism whereby foreign gene(s) are moved into an organism and inserted into that organism's genome. Retroviruses such as HIV serve as vectors by inserting genetic information (DNA) into the genome of human cells. Bacteria can serve as vectors in plant populations.
virus: viruses consist of a piece of nucleic acid covered by protein. Viruses can only reproduce by infecting a cell and using the cell's mechanisms for self-replication. They can cause disease; modified viruses can also be used as a tool in gene therapy to introduce new DNA into a cell's genome.
volunteer: in farming, a crop plant of one type that grows spontaneously within subsequent crops of a different type.

X

X chromosome: one of the two sex chromosomes. Females normally carry two X chromosomes. Compare Y chromosome.
xenograft: tissue or organs from an individual of one species transplanted into or grafted onto an organism of another species, genus, or family. A common example is the use of pig heart valves in humans.
xenotransplant: transplantation of tissue or organs between organisms of different species, genus, or family. A common example is the use of pig heart valves in humans.

Y

Y chromosome: one of the two sex chromosomes. Males normally carry one X chromosome and one Y chromosome.
yeast artificial chromosome (YAC): a vector used to clone DNA fragments (up to 400 kb); it is constructed from the telomeric, centromeric, and replication origin sequences needed for replication in yeast cells. Compare cloning vector.

Z

zygote: a cell produced by the fusion of a female gamete (egg cell or ovum) with a male gamete (sperm cell or pollen grain). The joining of a sperm and egg cell is called fertilisation. Zygotes are diploid and undergo cell division to become an embryo.

C

D

E

G

H

I

J

K

L

M

N

O

P

R

S

T

U

V

W

X

Z